Participatory Research in More-than-Human Worlds

Socio-environmental crises are currently transforming the conditions for life on this planet, from climate change, to resource depletion, biodiversity loss and long-term pollutants. The vast scale of these changes, affecting land, sea and air, has prompted calls for the 'ecologicalisation' of knowledge.

This book adopts a much needed 'more-than-human' framework to grasp these complexities and challenges. It contains multidisciplinary insights and diverse methodological approaches to question how to revise, reshape and invent methods in order to work with nonhumans in participatory ways. The book offers a framework for thinking critically about the promises and potentialities of participation from within a more-than-human paradigm, and opens up trajectories for its future development. It will be of interest to those working in the environmental humanities, animal studies, science and technology studies, ecology and anthropology.

Michelle Bastian is a Chancellor's Fellow in the Edinburgh School of Architecture and Landscape Architecture at the University of Edinburgh, UK.

Owain Jones is Professor of Environmental Humanities, School of Humanities and Cultural Industries, University of Bath Spa, UK.

Niamh Moore is a Chancellor's Fellow in the School of Social and Political Science at the University of Edinburgh, UK.

Emma Roe is Lecturer in Human Geography, University of Southampton, UK.

Routledge Studies in Human Geography

This series provides a forum for innovative, vibrant, and critical debate within Human Geography. Titles will reflect the wealth of research which is taking place in this diverse and ever-expanding field. Contributions will be drawn from the main sub-disciplines and from innovative areas of work which have no particular sub-disciplinary allegiances.

For a full list of titles in this series, please visit www.routledge.com/series/SE0514

61 Releasing the Commons
Rethinking the futures of the commons
Edited by Ash Amin and Philip Howell

62 The Geography of Names
Indigenous to post-foundational
Gwilym Lucas Eades

63 Migration Borders Freedom
Harald Bauder

64 Communications/Media/Geographies
Paul C. Adams, Julie Cupples, Kevin Glynn, André Jansson and Shaun Moores

65 Public Urban Space, Gender and Segregation
Women-only urban parks in Iran
Reza Arjmand

66 Island Geographies
Essays and conversations
Edited by Elaine Stratford

67 Participatory Research in More-than-Human Worlds
Edited by Michelle Bastian, Owain Jones, Niamh Moore and Emma Roe

Participatory Research in More-than-Human Worlds

Edited by Michelle Bastian, Owain Jones, Niamh Moore and Emma Roe

LONDON AND NEW YORK

First published 2017 by Routledge

2 Park Square, Milton Park, Abingdon, Oxfordshire OX14 4RN
52 Vanderbilt Avenue, New York, NY 10017

Routledge is an imprint of the Taylor & Francis Group, an informa business

First issued in paperback 2018

British Library Cataloguing in Publication Data
A catalogue record for this book is available from the British Library

Library of Congress Cataloging in Publication Data
Names: Bastian, Michelle, editor.
Title: Participatory research in more-than-human worlds / edited by Michelle Bastian, Owain Jones, Niamh Moore and Emma Roe.
Description: Abingdon, Oxon ; New York, NY : Routledge, 2017. | Series: Routledge studies in human geography ; 67 | Includes bibliographical references.
Identifiers: LCCN 2016027203 | ISBN 9781138957350 (hardback) | ISBN 9781317340881 (adobe reader) | ISBN 9781317340874 (epub) | ISBN 9781317340867 (mobipocket)
Subjects: LCSH: Human ecology—Research—Methodology. | Natural history—Research—Methodology.
Classification: LCC GF26 .P37 2017 | DDC 304.2072—dc23
LC record available at https://lccn.loc.gov/2016027203

ISBN: 978-1-138-95735-0 (hbk)
ISBN: 978-0-367-13874-5 (pbk)

Typeset in Times New Roman
by Apex CoVantage, LLC

Contents

Figures

Contributors

Jennifer Atchison is Senior Lecturer in Social Science and researcher at the Australian Centre for Cultural Environmental Research (AUSCCER), University of Wollongong, Australia. Her research focuses on the material interactions of human-nature relationships, including the ways in which nonhumans (particularly plants, but also others) shape human lives and provoke us to think differently and live more sustainably in the world.

Michelle Bastian is a Chancellor's Fellow in the Edinburgh College of Art at the University of Edinburgh. Her work focuses on the role of time in social practices of inclusion and exclusion. She was the Principal Investigator on an Arts and Humanities Research Council (AHRC) funded project that explored the possibility of extending participatory approaches to more-than-human communities, as well as a number of other AHRC-funded participatory projects. She has written on the inter-relations of time and environment and is a member of the Extinction Studies Working Group where she has extended her approach to look at the temporalities of species extinction.

Reiko Goto Collins and Timothy Martin Collins are environmental artists, researchers and authors working together since they first met in San Francisco in 1985. Their work is installation based, resulting in exhibitions, publications and public art opportunities. Over the last 15 years they have sustained an international art practice that is focused on the cultural (aesthetic, ethical and spiritual) elements of environmental change, referencing art history and the philosophical and scientific ideas about nature. Their focus over the last five years has been Scottish forests and landscapes. Working across artist materials, new media, philosophy, science and technology, they explore meaning through experience, reflection and in-studio practice. They are interested in the small contributions art makes to public discourse about places and things.

Eva Giraud is a lecturer in Media, Communication and Culture at Keele University. Her work explores tensions between activism, new materialism(s) and environmental politics.

Lesley Head is Redmond Barry Distinguished Professor and Head of the School of Geography at the University of Melbourne, Australia. Her research has examined human-environment interactions over space and time, with particular attention to plants.

Deirdre Heddon is Professor of Contemporary Performance at the University of Glasgow. She is the author of *Autobiography and Performance* (Palgrave Macmillan, 2008), co-author of *Devising Performance: A Critical History* (Palgrave Macmillan, 2005), and co-editor of a number of anthologies, including most recently *Histories and Practices of Live Art* (Palgrave Macmillan, 2012) and *It's All Allowed: The Performances of Adrian Howells* (Intellect, 2016). Her research has been published in various journals and editions, from Performance Research to Cultural Geographies, and emerges also in practice-based outputs (The Walking Library and 40 Walks). Dee has just launched a new series with Palgrave Macmillan, *Performing Landscapes*, for which she is writing *Performing Landscapes: Forests*.

Hester is a black and white spaniel. She is an expert small-mammal tracker and part-time research assistant with no formal institutional ties. Her previous projects include surveying for endangered mammals in Wales, and she is currently engaged in a long-duration ethological study of grey squirrels and feral pigeons.

Timothy Hodgetts is a Lecturer in Human Geography at Jesus College, Oxford; and a postdoctoral researcher at Oxford University's School of Geography and the Environment. His work centres on the multi-species modes of connectivity through which people live with more-than-human lives – from the big mammals that are 'like us', to the microscopic microbial communities that are 'in and around us'.

Gregory Hollin is a postdoctoral fellow at the School of Sociology and Social Policy, University of Leeds. His work is broadly concerned with the sociology of science and medicine.

Owain Jones is Professor of Environmental Humanities, University of Bath Spa. Owain specialises in geographies of nature-society relations, place and landscape, community and resilience, the role of memory, nonhuman agencies and temporalities of landscape. He is currently researching into water and community issues; tidal/coastal landscapes; co-production of knowledge with nonhumans; and community, memory, flooding and resilience in a series of AHRC and Economic and Social Research Council (ESRC) funded projects. He has an arts background and frequently collaborates with artists and works in interdisciplinary research teams. He has published a number of peer-reviewed papers on the above subjects, and has recently edited the book *Geography and Memory: Identity, Place and Belonging* for the Palgrave Memory Studies Series. He is currently supervising 3 Ph.Ds.

Anna Krzywoszynska is a Leverhulme Early Career Research Fellow at the Department of Geography, University of Sheffield. She is an interdisciplinary social scientist who uses qualitative research methods to understand the relationships between humans and nature, and the role scientific and other forms of knowledge play in these relationships. Her current research explores soil ontologies and their impact on agricultural and environmental futures.

Antony Lyons is an independent artist-researcher. With a background in eco/geo-sciences and landscape design, much of Antony's work is concerned with relationships between ecological processes, environmental change and social adaptation. Areas of particular focus include coastal/river landscapes, deep-time (geological) perspectives, routes/journeys and intangible cultural themes. His research and production methods rely on geopoetic creative fieldwork and experimental remixing of archives, recordings, data and contemporary narratives – explored in the context of both 'slow' and 'intensive' artist-residencies. Concerned with places and their possibilities, his methods include sculpture, film, sound and intermedia installation – addressing tensions, traces, transitions and environmental justice.

Jon Pigott (Cardiff Metropolitan University) Jon's artistic research activities currently involve making kinetic sound sculptures that explore the relationship between objects, systems and sound. His specialisms relating to this practice include histories and theories of sound art, sculpture and technology as well as various making processes such as digital fabrication and handmade electronics. Following a 12-year stint in the music industry helping to bring many high-profile music and film productions to fruition, Jon joined Cardiff Metropolitan University in 2008 where he currently teaches within the school of art and design.

Clara Mancini is a Senior Lecturer in Interaction Design at The Open University's School of Computing and Communications. She founded and heads The Open University's Animal–Computer Interaction Laboratory, whose mission is to expand the boundaries of interaction design and the design of interactive systems beyond the human species, thus contributing to (other-than-human and human) animal wellbeing, social inclusion, interspecies cooperation and environmental restoration.

Niamh Moore is a Chancellor's Fellow in Sociology at the University of Edinburgh. She has a background in interdisciplinary feminist studies, including over 10 years working as a participatory researcher, with projects including community food growing and community archiving. She has written *The Changing Nature of Eco/feminism: Telling Stories from Clayoquot Sound* (2015) about ecofeminist activism against logging on the west coast of Canada, and is creating an online archive of oral histories and other data from the research project. She is also currently working on a number of projects including on sustainable transport and mobilities, the Alexander Technique, archiving of research data and community-produced data, and action research with students. She teaches on ecofeminism, food, sustainability and research methods.

Hannah Pitt is a researcher at the Sustainable Places Research Institute, Cardiff University. Her work considers community gardens as more-than-human spaces of care, the effectiveness of social initiatives for food sustainability and blue-green spaces as sites of wellbeing. As a cultural geographer she takes a critical perspective on everyday interactions with nature, and works closely with third-sector partners.

Peter Reason is a writer, focusing in particular on ecological literature responding to the crisis in ecology. His award-winning book *Spindrift: A Wilderness Pilgrimage at Sea* (Vala Publishing Cooperative, 2014) weaves an exploration of the human place in the ecology of the planet into the story of a sailing voyage. Prior to his retirement from academia, he made major contributions to the theory and practice of action research in writing, teaching and research in the field of sustainability. He is Professor Emeritus at the University of Bath and Adjunct Professor in the Department of Transformative Studies at California Institute for Integral Studies. He blogs at onthewesternedge.wordpress.com and peterreason.net.

Emma Roe is a Lecturer in Human Geography at the University of Southampton. She has been working as a Nonhuman geographer for over 15 years developing novel methodologies and conceptual frameworks for exploring nonhuman agencies in the fields of embodied food consumption practices, food animal and laboratory animal welfare and, most recently, antimicrobial resistance and infection prevention. ESRC, AHRC, EPSRC, the European Commission, British Academy and Wellcome Trust funded this research. She has written over 30 peer-reviewed journal articles, book-chapters, reports and short articles on food consumption practices, the market for animal welfare-friendly foodstuffs, ethical responses to sentient creatures, ecological citizenship and experimental research methods. She has presented research findings to industry, policy and academic audiences, nationally and internationally. Currently she supervises three PhD students.

Hollis Taylor is a violinist/composer, zoömusicologist, ornithologist and Research Fellow at Macquarie University. She previously held research fellowships at the Institute for Advanced Study in Berlin, the Muséum national d'Histoire naturelle in Paris and the University of Technology Sydney. Taylor has an abiding interest in animal aesthetics, particularly vis-à-vis Australian songbirds. She performs her award-winning (re)compositions of pied butcherbird songs on violin along with various outback field recordings, and her creative practice also takes in sound and radiophonic arts. In *Post Impressions: A Travel Book for Tragic Intellectuals*, Taylor documents (in text, audio and video) Jon Rose and her bowing fences throughout Australia, and her monograph, *Is Birdsong Music? Outback Encounters with an Australian Songbird*, is forthcoming. She is webmaster of www.zoömusicology.com.

Acknowledgements

We are grateful for the following permissions to reproduce previously published work:

to the Association for Computing Machinery (ACM) for permission to reproduce the article; Mancini, C., 2011. Animal–computer interaction: a manifesto. *Interactions*, 18 (4), 69–73.

to Elsevier for permission to reproduce sections from the article; Mancini, C., 2016. Towards an animal-centred ethics for animal–computer interaction. *International Journal of Human-Computer Studies*, online first, 25 April.

Introduction

More-than-human participatory research

Contexts, challenges, possibilities

Michelle Bastian, Owain Jones, Niamh Moore and Emma Roe

This collection arises from an AHRC-funded research project called *In Conversation with. . .: codesign with more-than-human communities* that ran in 2013, as well as a series of panels held at the RGS-IBG International Conference in 2014 on the *Co-Production of knowledge with non-humans*. In both cases we sought to explore the notion of a 'more-than-human participatory research'. Yet to say 'more-than-human participatory research' seems like too much of a mouthful. These are words that do not roll easily off the tongue, but instead suggest some kind of cacophony, some noisy dissonance. These are words that seem like they should not really sit beside each other, words that do not quite make sense.

Nonetheless, our aim in this collection, which we will explain further below, is precisely to explore the potential of bringing together the growing field of 'more-than-human research' (MtHR) with the more established practices of 'participatory research' (PR). In bringing these seemingly disparate fields together, we want to point to more entwined histories than initially might seem obvious, and at the same time, to also open up a series of new questions: What might it mean to invite 'the more-than-human' to be an active participant, and even partner, in research? How are prevailing ways of conceiving research in terms of issues of knowledge, ethics, consent and anonymity challenged and transformed when we think of the more-than-human as a partner in research? How might it be possible to transform existing frameworks, practices and approaches to research? What would this transformed research look like?

We first situate more-than-human participatory research (MtH-PR, to help, perhaps, with the cacophony?) within a context of socio-environmental crisis. As we write, the two great conjoined 'issues' of shared planetary life – social and ecological injustice (flagged up by the Brandt Report in 1980 and repeatedly after) – seem to be entering new levels of starkness and volatility. These crises are 'headlined' by climate change, but also include resource depletion, biodiversity loss, and long-term pollutants among others. Attention has also been called to the uneven ways that the consequences of living in this changing world are felt and experienced by specific humans and nonhumans. The vast scale of these changes, which are having profound effects on communities living on land, in the sea and air, have prompted calls for the 'ecologicalisation' of knowledge as an essential step in moving away from Enlightenment philosophies of rational, self-aware humans in a machine-like world (Plumwood 2002, Latour and Weibel 2005, Code 2006, Hinchliffe 2007).

Thus we also situate MtH-PR in the context of widespread experimentation with methods, and related rethinking of methodology, in the social sciences and beyond. We are hearing of inventive methods (Lury and Wakeford 2013), live methods (Back and Puwar 2012), mobile methods (Büscher *et al.* 2010), materialist methods (Pryke et al. 2003), creative methods (Gauntlett 2011), mixed methods (Brannen 2005) and methods for working with big data (Savage and Burrows 2007). While this interest in methodological innovation has been linked with new funding contexts, and demands for novelty, as well as calls for greater accounting of research impact, Wiles *et al.* (2010, p. 11) more generously recognise that there are other impetuses for innovation including theoretical, ethical and practical motivations. Our efforts to imagine the possibilities of MtH-PR is thus driven by the need to take environmental devastation seriously, and to develop research methods that might better support more sustainable ways of living together.

Future directions for more-than-human research methods

At the heart of much of this methodological experimentation is the conviction, which was at the heart of earlier feminist interventions into methods debates (Harding 1986, 1987, Haraway 1988), that methods don't just describe worlds, but make worlds (Law 2004). That is, they make some things more visible and others more difficult to take into account. As a result, research on aspects of social life that have been absent from dominant research paradigms has brought with it a multitude of critiques of dominant research methods and the search for new methods and new ways of working with traditional ones.

The world of what might be broadly termed more-than-human research (e.g. animal geographies, critical animal studies, ecofeminism, environmental humanities, human–animal studies, multi-species research, new materialism, queer ecologies, science and technology studies [STS], etc.) has been no different. This is research that has sought – in one way or another – to take nonhuman life, and the entanglements of human/nonhuman life, seriously and to thus step away from the modernist dismissal of nature and nonhumans as anything but resources. For those working in these and related areas, questioning the methods by which knowledge is created, and science is 'done', is key to shifting away from paradigms of human exceptionalism. As a result, here too we see methods being augmented, hybridised and remade. Examples include etho-ethnology and ethno-ethology (Lestel et al. 2006), multi-species ethnography (Kirksey and Helmreich 2010) as well as those methods adopted for use within zoömusicology (Taylor 2013), animal–computer interaction (Mancini, in this volume) and animal geographies (Wolch and Emel 1998).

Despite, or perhaps because of, the fledgling character of many more-than-human research methods there have already been a number of literature reviews that have sought to trace out the territory and offer suggestions for ways forward. This includes a review of multi-species ethnography (Ogden *et al.* 2013), as well as two reviews of methods within animal geographies (Buller 2014, Hodgetts and Lorimer 2014). In each there is an underlying concern with decentring the human and with taking nonhumans' experiences, perspectives and agencies seriously, in ways that are situated, embodied and non-homogenising. Thus Henry Buller hopes that animal geography will develop approaches that are able to 'suggest or reveal what matters, or what might matter, to animals as subjective selves' (2014, p. 7), while Timothy Hodgetts and Jamie Lorimer emphasise the importance of fulfilling animal geography's 'promise of taking animals seriously as subjects and ecological agents' (2014, p. 8).

All also emphasise the positive possibilities of working critically with scientific knowledge, technologies and methods as part of achieving these aims. For example, when setting out the future direction of animal geography methods, Buller (2014, p. 7) proposes that a greater engagement with the biological and animal sciences in particular will be needed. For Hodgetts and Lorimer, key suggestions are technologies that enable the monitoring, tracking and analysis of animals spatial movements, experiments with intra- and interspecies communication, and genomic methods that give insights both to 'historic animal mobilities' and 'microbial ecologies within and between animal bodies' (Hodgetts and Lorimer 2014, p. 3). Yet, if the aim is to 'suggest or reveal what matters' to nonhumans, then another contribution to this methodological bricolage might come from a quite different approach, specifically methods developed by colleagues in participatory geographies as well as PR more generally. As an area that is focused on the inclusion of marginalised voices and experiences, the subversion of dominant power structures and has a commitment to co-producing research with those who are affected by it, there appears much to be gained by including it in the conversation.

Potential affinities between MtHR and PR

Questions over the relevance of academic research for broader constituencies have led to increasing interest in PR practices and their overarching aspiration of developing socially responsible and democratic research methods. Such approaches have turned to the co-production of knowledge as a way of transforming the power relations, goal-setting methods and expected outcomes of the research process. Central components of this agenda have been the desire to support the inclusion of marginalised actors and to make research accountable to those it affects. PR has also had a long history of grappling with problems around who is understood 'to know' within the research process. Methods have been developed in order to challenge what kinds of knowledges are seen to be legitimate, while also attending to the problems of producing knowledge within contexts of stubborn inequality. The aim has been to decentralise knowledge creation, and question the legitimation of knowledge by 'experts' operating outside of research subjects' subjective experience, through moving towards a distributed democratic, transparent process that also provides a new route for addressing social justice.

While PR methods have been principally concerned with the exclusion of particular human communities, there have been calls from within the area to respond to a further, often unacknowledged, exclusion of the more-than-human. Participatory action researcher Peter Reason (2005), for example, has argued that the more-than-human is the cutting edge problem for PR in the context of the Anthropocene. While participatory economic geographers J.K. Gibson-Graham (2011) and Gerda Roelvink (Gibson-Graham and Roelvink 2010) have called for the extension of work around community economies to more-than-human collectives, arguing for the importance of reframing research as 'a process of learning involving a collective of human and more-than-human actants – a process of co-transformation that re/constitutes the world' (Gibson-Graham and Roelvink 2010, p. 342). Other examples include Kye Askins and Rachel Pain's exploration of the role of materiality in PR and particularly the way that 'objects as conduits may facilitate transformative social relations to seep across spaces of encounter' (2011, p. 817, see also Roe and Buser 2016). However, as Isabelle Stengers (2015) argues, one of the great failings of recent political and knowledge cultures was that:

> [our generation] thirty years ago, participated in, or impotently witnessed, the failure of the encounter between two movements that could, together, perhaps have created the political intelligence necessary to the development of an efficacious culture of struggle – those who denounced the ravaging of nature and those who combated the exploitation of humans.
>
> (2015, p. 10)

Ecofeminists did try to do make these connections, insisting on the inseparability of struggles for nature and for social justice, but were roundly critiqued for universalism and essentialism – essentialism being the term used then for making sure that matter was made not to matter (Moore 2015, pp. 216–230). So it is interesting to see some of this work surface in more recent discussions. For instance, Carol Adams's (1990) work is now more widely being taken up in critical animal studies. Other classic ecofeminist texts such as Susan Griffin's *Woman and Nature: The Roaring Inside Her* (1978) may also acquire new resonances in current times. The epigraph to Griffin's book, for example, reads: 'These words are written for those of us whose language is not heard, whose words have been stolen or erased, those robbed of language, who are called voiceless or mute, even the earthworms, even the shellfish and the sponges, for those of us who speak our own language' (1978, v). It could also stand as an epigraph for this collection, particularly in her call for taking seriously the task of listening to and working with the more-than-human.

Even still, the challenges of bringing movements together need to be acknowledged and Anna Tsing offers one reflection on this in her discussion of collaborations between environmentalists and indigenous peoples. Tsing's response to those who have understood environmentalists' interest in indigenous knowledge 'only as a repetition of environmentalists' fantasies and imperial histories' is to lament the persistence of familiar metanarratives 'in which nothing good can happen – good or bad – but more of the same' (2005, p. 4). She turns to 'friction' to suggest 'the awkward, unequal, unstable, and creative qualities of interconnections across difference' (2005). We argue that bringing together PR and MtH offers one means of, albeit belatedly, developing such an 'efficacious culture of struggle' (Stengers 2015, p. 10), one where frictions appear as generative.

Thus, in proposing a move towards an MtH-PR, we want to recognise these difficulties, while also suggesting that there are a range of intriguing overlaps between the commitments of PR and many MtH approaches. For example, both are interested in developing methods that can reveal what matters to those traditionally excluded from dominant knowledge making processes, as well as fostering techniques that challenge hierarchies in the hope of 'creating with' in ways that are ethical, socially just and epistemologically open. As a result, we would argue that an engagement with the various debates that have taken place within PR offer a rich opportunity for those working with nonhuman others to reflect on their methodologies in complex and sophisticated ways. Further, PR may also benefit given moves towards a more explicit recognition of the participation of the more-than-human in collaborative research.

Diverging co-productions

A further example that at first glance seems to suggest important affinities between MtHR and PR is hinted at via scattered references throughout Buller's review, in particular, but also in Ogden *et al.* That is, the use of terms such as 'participatory' (Buller 2014, p. 4), 'co-creation' (Buller 2014, p. 6), 'co-production' (Buller 2015, p. 6) and 'coproductionist framework' (Ogden et al. 2013, p. 12). However, as became evident in the process of developing this collection, bringing together PR and MtH frameworks highlighted a more general need to pay explicit attention to the different histories of the term 'co-production' arising from each research area. That is, for MtH researchers (particularly those working within or inspired by STS) co-production seems to most often refer to the more general idea that human and nonhuman agents are intertwined in shared worlds, with both involved in the 'production' of these worlds. This approach emphasises a questioning of nature/culture divides and the disciplinary divides based on them. For PR, however, co-production more often focuses attention on efforts to subvert the divide between researcher and researcher, in order to move from research *on* to research *with*. That is, while in the former, co-production offers an analytical framework for approaching the object of study, in the latter, co-production is a method of engaging with fellow enquirers.

A helpful way of demonstrating this distinction (which is, of course, broad brush as most such distinctions are), is through a comparison of the ways that coproduction is used and defined in the work of STS theorist Sheila Jasanoff (2004) and political scientist Elinor Ostrom (1996). As discussed by Jennifer Atchison and Lesley Head in this volume, Jasanoff uses the term co-production to emphasise the need to think the natural and the social together (2004, p. 4). Her emphasis is on how knowledge of the natural and the social world is produced, and particularly the claim that scientific (and technological knowledge) 'is not a transcendent mirror of reality. It both embeds and is embedded in social practices, identities, norms, conventions, discourses, instruments and institutions – in short, in all the building blocks of what we term the social' (Jasanoff 2004, p. 3). She further describes co-production explicitly as 'an interpretive framework' (2004, p. 6).

This emphasis on co-production as an analytical tool is important because, by contrast, for Ostrom co-production is better understood as a 'process' (1996, p. 1073). The targets of critique, for her and her colleagues, were theories of public governance that supported widespread centralisation of services (1996, p. 1079). They argued, instead, that 'the production of a service, as contrasted to a good, was difficult without the active participation of those supposedly receiving the service' (1996, p. 1079). Importantly, this participation is directly set against 'citizen "participation" in petitioning others to provide goods for them' (1996, p. 1083), with the scare quotes suggesting an emphasis on a more active engagement in the process. Further, the more general empowerment of participants in the process is also valued, with reports 'that local activism through coproduction rapidly spills over to other areas' used to suggest added benefits of the approach (1996, p. 1083). In this case then, co-production is more closely linked with active processes of engaging with, and empowering, those involved.

One needs to be cautious then, when, suggesting that the use of terms such as participation, co-production and co-creation within MtHR might suggest an inclination towards the possibility of an MtH-PR. That is, given the different understandings set out here, and the confusions we experienced when we framed this project as more-than-human coproduction (as we did at the RGS-IBG in 2014), we want to insist on the specificity of PR accounts of participation and coproduction. Like MtHR, these accounts acknowledge that the world is 'co-shaped' by multiple actors. However, they provide a specific emphasis on the processes by which these actors can become actively engaged in research in order to develop responses to specific issues they are facing. Further we want to insist that the provocation of this collection is to explore this latter account of coproduction. That is, could MtH commitments to understanding 'what matters' to nonhumans support even more challenging methodological experiments, particularly around who research is done for and with?

As a result, whilst there has been a steady growth in work that recognises the agency of nonhumans in knowledge production, something different characterises the contributions to this edited collection and that is an interest in how one might invite specific nonhumans into the research process at the outset, rather than identifying nonhuman agency in human social worlds as a research output. This collection builds on the wide range of work that challenges the Western heritage of machine-like understandings of animal nonhumans (or inert-matter for non-animal nonhumans) by exploring what the next steps might be in terms of academic research practices. To date, PR and MtH trajectories have not yet been brought into explicit conversation; however, each appears to have much to learn from the other. This collection thus presents research from a wide range of disciplines, regions and methodological approaches that grapple with the problem of how to revise, reshape and invent methods in order to work with nonhumans in participatory ways. The challenges are considerable, and yet interest in this area is intensifying. This collection therefore offers an initial framework for thinking critically about the promises and potentialities of participation from within a more-than-human paradigm, and opens up trajectories for its future development.

The revenant of anthropomorphism

Before discussing the individual contributions, we want to address the (almost inevitable) question of what role anthropomorphism might play in this venture. As may be familiar to our readers, many of those who work in the broad area of MtHR can end up repeatedly having to argue for the possibility of relationships between humans and nonhumans that go beyond the purely instrumental, or to respond to generalised critiques that they are anthropomorphising (i.e. critiques made, not in relation to the specifics of the work, but simply because someone is talking about nonhuman agencies). Indeed the situation reminds us of similar situations that arise within feminist research where there can be pressure to return to foundational questions such as whether there is indeed any problem with sexism anymore. Within feminism there has been recognition that this demand can hinder feminist work by taking energy away from developing and deepening feminist theory because basic assumptions have to be proven and reproven. We see the possibility of an analogous problem arising within MtHR. As a result, we were keen that this collection built on previous work to push questions of MtHR and any potential interrelations with PR further, rather than revisited an issue that has been tackled directly elsewhere. Thus while we will address responses to anthropomorphism here, this is in part to free up our contributors from the obligation so that they can explore the specifics of their case studies in relation to the collection's frame of MtH-PR. This is not to say that there is no need to be careful of bias, inappropriate assumptions or projection, but to suggest that these are problems that all research methods are developed to grapple with, even those focused on humans.

One response to anxieties about anthropomorphism is to point to the huge amount of research that has shown that nonhumans are capable of a much wider range of cognitive, emotional and symbolic behaviours than they have traditionally been given credit for in Western cultures. Wolves and dogs have senses of fairness and justice (Bekoff 2007), parrots call each other by name (Berg 2011), octopuses use tools (Finn et al. 2009) and mimosa plants can learn to distinguish between types of threats (Gagliano et al. 2014). As Bastian points out in her chapter in this volume, PR has long been wary of claims of deficits in ability and instead emphasises methodological flexibility and experimentation in order to find ways of including all those affected by an issue. The growing awareness of the wide range of capacities that nonhumans enjoy, suggests that an MtH-PR could find ways of working with these capacities, rather than assuming from the outset that such research was impossible.

Another response is to reject the premise that is fundamental to generalised accusations of anthropomorphism, namely human exceptionalism. As philosopher Val Plumwood (2007) has argued, in her blistering critique of Raymond Gaita's (2002) *The Philosopher's Dog*, the belief in a hyperseparation between human and nonhuman animals leads to untested assumptions of radical discontinuities, and a lack of curiosity about evidence that might prove the contrary. For Plumwood, the real problem is not anthropomorphism, but with 'the way assumptions of human superiority and mind discontinuity structure our concepts and limit our perceptions of animal behaviour' (2007, n.p.). This scepticism over the ability of nonhumans to have any mindful or communicative capacities, is not, she further argues, 'purely an empirical or observational matter, but is always already an action of exchange or refusal of exchange, a matter of stance and performativity (in the sense of Wittgenstein and Austin), a matter of listening and invitation' (2007, n.p.). Plumwood highlights the political nature of this performative scepticism through comparison with contexts where other humans have been thought to be without 'proper' reflective capacities such as slavery and colonialism (2007, n.p.). Attention might also be given to how the dominant figure of the *anthropos* continues to illustrate the 'inability of Western knowledges to conceive their own processes of (material) production, processes that simultaneously rely on and disavow the role of the body' (Grosz 1993, p. 187). In contrast then, a range of theorists have suggested other terms that might be more useful in drawing attention to inappropriate assumptions about nonhumans including Eileen Crist's (1999) 'mechanomorphism' (the assumption that animals are like machines) and Daniel Dennett's (cited in Pollan 2013) 'cerebrocentrism' (the assumption that only a biological brain can support intelligence). Or following Grosz, to take seriously the corporeality of the human body, as a basis for relating to and sharing experiences with nonhumans, for example suffering (Haraway 2008).

Of course we are not claiming that therefore PR with nonhumans would be straightforward and unproblematic. PR rests in large part on careful, systematic, ethical listening, conversations, non-specialist languages, trying to establish nonhierarchical power relations, conducting research in conducive settings and material arrangements. To do all those with nonhumans, of one stripe or another, raises a whole suite of conceptual, ethical and practical challenges. We would suggest however that, when keeping in mind the very small amount of scientific research on the full capacities of specific nonhumans, as well as the general prejudices within heritages of Western thought,[1] there is a lot of room to be curious about how these challenges might be met.

What we aren't saying

We also thought it important to clarify what we *don't* think this collection is doing. That is, our aim is not to set out an already developed set of methods for MtH-PR. While we do include examples of researchers actively wrestling with the possibility, there is still much work to be done. Thus our collection is motivated by what we see as a highly promising potential for more-than-human researchers and participatory researchers (who are not always different people), to explore how their shared concerns with including those who have been excluded in particular ways from research processes, might speak to each other. These explorations may very well create challenges to, just as much as providing new ways of supporting, each other's ways of working. Thus, we seek to contribute to broader methodological discussions occurring in these areas, including around the theoretical questions underpinning the choice of particular methods, such as whom these methods include/exclude, where they do and don't work, what kinds of ethical/relational considerations they raise, what are the frictions and affordances and how they might be imagined otherwise.

The step between recognising the agential capacities of specific nonhumans, to then developing methods that might enable their active participation in research processes, is a significant threshold, one that this current collection identifies and seeks to begin to cross. Articulating and unpacking some of the difficulties that might be encountered is thus a key contribution. Given the range of nonhuman others we share the planet with (and who are present in this book), as well as the fledgling nature of this field, we do not seek to propose a systematic or unified approach, but rather to introduce readers to a range of work unfolding in this area and a set of interconnecting themes and questions.

The question of what the limits of an MtH-PR might be is also a live issue throughout the chapters (see particularly 10–12). For example, could specific nonhumans ever 'fully participate' in a research project, particularly when we consider the PR ideal of participants being able to actively shape research questions, the processes of data gathering and analysis and dissemination strategies? Here we would note, however, that the question of full participation is still a live one within PR with humans,[2] and so we have no expectation that such questions will be easily answered. However we do believe that they should indeed be asked in relation to research with nonhumans, as part of responding to the ethical framework underlying PR, namely that those affected by research should be involved in it. So, while the aim of the book is to showcase work developing in this area, it is inevitable that as many questions will be raised as answered. No more so than in the specifics of how exactly such participation might occur and in what ways. As such the wide range of case studies enables the collection to offer insights into what these specifics might eventually look like in relation to different nonhuman partners and contexts.

What we *are* hoping to do is to support a more explicit dialogue between PR and MtHR via the provocations suggested in this collection. We see strong resonances between the more-than-human and participatory paradigms, but we are not claiming to know in advance how the dialogue between them would play out. Instead the collection aims to take both as seriously as possible and to explore the affordances and frictions. Where do analogies offer new insights and where do they break down? We by no means assume that one can be laid over the other, but rather that the perspectives of each might allow the other to be seen in a different light (see for example Bastian's discussion of diffraction, in this volume). For us, MtH-PR arguably goes beyond providing a new context for research (Wiles *et al.*

2013), and the demands of trying to take the more-than-human seriously as a research participant call for a significant transformation in research methodology.

The book is divided into three sections. The first 'Experiments in more-than-human participatory research', the second 'Building (tentative) affinities' and the third 'Cautions'. These relate to how the chapters make different forms of critical engagements with doing MtH-PR.

Review of chapters

The first section 'Experiments in more-than-human participatory research' includes chapters that make the boldest moves for putting MtH-PR into practice. In the opening chapter of the collection, Michelle Bastian describes and analyses a one-year exploratory project that speculated about the possibilities of working with animals, insects, plants and the elements, as research partners. She proposes reading PR and MtHR through a diffractive lens to see what light each might shed on the other. Drawing on a recent review of PR as well as critiques, such as Bill Cooke and Uma Kothari's (2001), Bastian suggests a number of ways that the approaches and insights developed within PR might usefully point the way towards a MtH-PR. She also shows how MtH approaches might encourage a reconsideration of particular aspects of PR. This includes issues such as a closer attention to the ways humans are shaped by the nonhumans in their life-worlds, whether PR approaches might help foreground and question power relationships between humans and nonhumans, how they also might help to challenge assumptions of competency, and encourage methodological exploration in order to support wider inclusions. However, the chapter also emphasises the dangers of assuming that participation is a simple good, and also explores issues of overlooking wider inequalities, the danger of pseudo-participation and the lack of wider contexts that might support working with nonhumans in these ways.

The focus of Hollis Taylor's chapter is birdsong, specifically Australian pied butcherbirds, and her groundbreaking work in zoömusicology. Arguing against the dominant approach of biologists studying a small number of songbird species in captivity, Taylor outlines a participatory ethnographic approach that works with free-living birds and which is led by musicologists. She reflects on her own practices by bringing them into conversation with participatory approaches, such as those particular to music including jazz and music therapy. Drawing on a variety of PR accounts, Taylor explores the problems of differences in skills and abilities, being both an insider and an outsider, how to ascertain if birds have given her permission to run the project with them, and coproducing research outcomes. She concludes by issuing a challenge to human exceptionalism within music and arguing instead for attending to nuance, individual capacities and the creativity of nonhuman songsters.

From music, we then turn to technology with Chapter Three including a selection of work from computer interaction designer Clara Mancini. In the first section, Mancini draws on the user-centred approach of Human-Computer Interaction (HCI) to offer an initial manifesto for an Animal–Computer Interaction (ACI) that designs technology with and for the nonhumans that are expected to work with it. Examples include in agriculture (where cows might interact with robotic milking machines) and in scientific research (where animals might be tagged and tracked with a variety of devices). She asks the question: How might the animal perspective inform the design of these and other technologies? She does this partly to mitigate risks of unsuitable designs, but more broadly in order to develop a research agenda around interspecies computer interaction. The second section brings us to more recent work in ACI which sets out an ethical framework for the approach. Here we focus particularly on Mancini's recommendations around animal welfare and animal consent in research processes, as key areas for consideration within a future MtH-PR. In particular Mancini argues that it in their specific role as users and participants that an ethical treatment of nonhuman animals should be considered.

Peter Reason's chapter takes us to the question of the 'participatory mind' and particularly how it might support a 'deep participation' with the more-than-human world. Working in the form of nature writing (arguably a genealogical progenitor of MtH-PR), Reason narrates a sailing journey off the northwest coast of Scotland. He focuses our attention on the temporality of pilgrimage in geological landscapes, and the intertwinings of clock time, Earth time, the eternal present and deep time. Playfully remaking the action research cycles of action and reflection within the experience of pilgrimage, Reason's contribution emphasises the role of writing forms in sharing participatory encounters more widely.

The second section 'Building (tentative) affinities' includes chapters that examine practices of learning to engage with animals, plants or water as research participants. Inviting another set of participatory literatures to the conversation, Timothy Hodgetts and his spaniel Hester take us into the world of wildlife conservation. In this chapter they share some of the ways they have learned to work together as part of a conservation team helping to spot traces of endangered pine martens. Attunement is suggested as a key issue for MtH-PR, particularly the ways that indexical communicative signs might be translated across species. They also raise questions about the levels of participation available to both human and dog within the endeavour, the uneven power between them and who counts within the larger conservation project. Finally, issues emphasised in PR literatures, namely ethical issues of consent, mutual benefit and recognition, are refracted through their experiences to highlight the difficulties of straightforwardly reading their activities, or participatory wildlife conservation more generally, as a potential form of MtH-PR.

The theme of apprenticeship continues in Hannah Pitt's contribution, but with the potential challenges extended by considering the possibilities of learning from plants. Building on her previous work on more-than-human methods, Pitt suggests that research conducted through processes of learning (e.g. via participatory action research, communities of practice, and/or apprenticeship), rather than through demonstrations of expertise, offers room for a greater recognition of 'planty knowledge'. The importance of nonverbal communication and active material engagement within apprenticeship, suggest it as a method that might support learning from plants through watching, growing, accepting feedback and trying other routes. Pitt emphasises, however, the otherness of plants and particularly that which must always remain elusive and obscure. While she notes this as a common problem with human PR, this elusiveness would appear to frustrate the future-looking aspects of PR that aim for shared visions of better worlds. She concludes by emphasising a range of 'tricky' problems that any MtH-PR with plants would face, including the difficulties of agreeing goals, aims and who is to be empowered in the process.

For Reiko Goto Collins and Timothy Martin Collins, moving to an understanding of plants as beings of value requires finding ways of supporting imagination and empathy. Their chapter describes their project 'Eden3: Plein Air', a sculptural instrument that a tree 'plays'. Temporality becomes an important issue here too, with the common assumption that trees are relatively still and passive in part resting on their different paces of growth and change. As part of counteracting these assumptions Goto Collins and Collins's project hones in on the processes of transpiration and photosynthesis which occur on a day to day level and can feel more similar to human time. Their chapter shares the difficult iterative process involved in working across art, science, technology and sound design in order to support an empathic relationship with a tree. They conclude by asking whether hearing the tree, its 'breath', might encourage a wider sense of ethical duty beyond the human. Here an MtH-PR is fundamentally about active listening and aiming to make a positive difference.

The collection's exploration of human interrelationships with plants continues in Anna Krzywoszynska's account of empowerment, skill and the creation of new subjectivities. She draws on her more-than-human ethnography of organic wine making to explore how each of these aspects of PR operate when read through the process of learning to care for vines. The importance of the relational self for empowerment is itself shown to be shaped by the affective states that enabled Krzywoszynska to move toward an active relationship with the vines she learned to prune. Enchantment, becoming and focus were all central to this skill acquisition. This analysis is then returned to PR debates on empowerment to show how a more-than-human perspective can offer further ways of understanding the process of cultivating new ways of being in the world.

The question of 'giving voice' is one that arises throughout this collection, and is a particular focus in Jon Pigott and Antony Lyons's contribution. Here the more-than-human participants include bats, the water in a river catchment, sensor technologies and data. Pigott and Lyons describe the development, and theoretical implications, of their eco-art project 'Shadows and Undercurrents'. With a primary interest in highlighting the hidden processes contributing to the loss of biodiversity, the project involved an 18-month 'slow-art residency' that culminated in an immersive installation space that included data-activated kinetic sculptures. Attunement, empathy and affect play alongside tools of measurement and the production of data-streams to produce an experience of 'intimate science'. Rather than seeking to facilitate the participation of more-than-humans from the outset, Pigott and Lyons reflect on the ways that their participation became more pronounced through the iterative development of the project. They conclude by focusing on water in particular, and speculating on ways that it may, and may not, be engaged with in participatory ways.

The final three chapters of the book, found within the third section 'Cautions', continue many of the themes raised throughout the collection, but also help to bring a particular focus on the potential limits of a MtH-PR. The role of empathy in developing greater recognition of nonhuman agency is emphasised in a number of contributions in this collection, as well as in MtHR more widely. In Eva Giraud and Gregory Hollin's chapter, however, the instrumentalisation of empathy is shown to be a key tool in the efficient handling of laboratory dogs, specifically beagles. Analysing documents arising from the experimental beagle colony at University of California, Davis, Giraud and Hollin show that care-taking practices and affective human–animal relations do not always generate the sense of ethical responsibility that MtHRers might hope it does. Here the differentiation between the two approaches to coproduction, discussed above, become of key importance. That is, while there is clear evidence that beagles coproduced research at Davis in the STS sense, any processes of coproduction, in the PR sense, were fundamentally undermined. Giraud and Hollin thus suggest that the complexities of resistance and consent need to be thoroughly engaged with, and particular attention paid to the ways violence can be intertwined with care.

Jennifer Atchison and Lesley Head's contribution reflects on their previous ethnobotanical research and emphasises the difficulties of considering plants as collaborators. The entanglements that they have studied between humans and plants demonstrate the possibilities of mutual flourishing, but also of brutally adversarial relationships, such as those with invasive plants. They also suggest that attending to any concerns that plants might have is not often seen to be a relevant or urgent task for researchers. Even if a MtH-PR were to be attempted with plants, the human framing of PR, the lack of knowledge of plant capacities, the need to attend to the specificities of particular plants as well as the significant methodological innovation that would be required all suggest reasons to be anxious and cautious about the endeavour. Tracing their work with yams, wheat and rubber vines, as well as the people entwined with them, Atchison and Head argue for more humble recognition of plants' diverse capacities and ways of being.

Deirdre Heddon's contribution concludes the collection by offering a critical response to the *In conversation with...* project discussed in Chapter One. Utilising a performative writing style, Heddon argues for closer attention to the importance of listening to *both* humans and nonhumans in any MtH-PR. Drawing on philosopher Jean-Luc Nancy's work, amongst others, Heddon reworks listening as concern, curiosity and anxiety. Introducing participatory forms of theatre-making, she proposes a form of co-authorship and collaboration that focuses on what can be made *with* others, through openness and acceptance, rather than traditional forms of academic scholarship that focus in individual demonstrations of expertise. She asks the questions: How might we avoid 'compelling the other to talk'? How might we avoid hearing only what we already know?

Notes

1 See also Kim Tallbear's (2011) argument about the ways these heritages continue to affect MtHR with the exclusion by some of nonliving nonhumans from accounts of agency.
2 See for example Cooke and Uma Kothari (2001), and the debates that followed the publication of this collection.

References

Adams, C. 1991. *The Sexual Politics of Meat.* New York: Continuum.
Askins, K., and Pain, R., 2011. Contact zones: participation, materiality, and the messiness of interaction. *Environment and Planning D: Society and Space*, 29 (5), 803–821.
Back, L., and Puwar, N., eds., 2012. *Live methods.* Malden, MA: Wiley-Blackwell.
Bekoff, M., 2007. *The emotional lives of animals.* Novato, CA: New World Library.
Berg, K.S., Delgado, S., Cortopassi, K.A., Beissinger, S.R., Bradbury, J.W., 2011. Vertical transmission of learned signatures in a wild parrot. *Proceedings of the Royal Society of London B: Biological Sciences*, 279 (1728), 585–591.
Brannen, J., 2005. Mixing methods: the entry of qualitative and quantitative approaches into the research process. *International Journal of Social Research Methodology*, 8 (3), 173–184.
Buller, H., 2015. Animal geographies II: methods. *Progress in Human Geography*, 39 (3), 374–384.
Büscher, M., Urry, J., and Witchger, K., eds., 2010. *Mobile methods.* Abingdon, Oxon: Routledge.
Code, L., 2006. *Ecological thinking: the politics of epistemic location.* Oxford: Oxford University Press.
Crist, E., 1999. *Images of animals: anthropocentrism and animal mind.* Philadelphia, PA: Temple University Press.
Finn, J.K., Tregenza, T., and Norman, M.D., 2009. Defensive tool use in a coconut-carrying octopus. *Current Biology*, 19 (23), R1069–R1070.
Gagliano, M., Renton, M., Depczynski, M., Mancuso, S., 2014. Experience teaches plants to learn faster and forget slower in environments where it matters. *Oecologia*, 175 (1), 63–72.
Gaita, R., 2002. *The philosopher's dog: friendships with animals.* Melbourne: Text Publishing Company.
Gauntlett, D., 2011. *Making is connecting: the social meaning of creativity, from DIY and knitting to YouTube and Web 2.0.* Cambridge, UK: Polity Press.
Gibson-Graham, J.K., 2011. A feminist project of belonging for the Anthropocene. *Gender, Place & Culture: A Journal of Feminist Geography*, 18 (1), 1–21.
Gibson-Graham, J.K., and Roelvink, G., 2010. An economic ethics for the Anthropocene. *Antipode*, 41 (s1), 320–346.
Griffin, S. 1978. *Woman and Nature: The Roaring Inside Her.* New York: Harper & Row.
Grosz, E., 1993. Bodies and knowledges: feminism and the crisis of reason. *In*: L. Alcoff and E. Potter, eds. *Feminist Epistemologies.* London: Routledge, 187–215.
Haraway, D., 1988. Situated knowledges: the science question in feminism and the privilege of partial perspectives. *Feminist Studies*, 14 (3), 575–599.
Haraway, D., 2008. *When species meet.* Minneapolis, MN: University of Minnesota Press.
Harding, S., 1986. *The science question in feminism.* Ithaca, NY: Cornell University Press.
Harding, S., ed., 1987. *Feminism and methodology: social science issues.* Bloomington, IN: Indiana University Press.

Hinchliffe, S., 2007. *Geographies of nature: societies, environments, ecologies*. London: Sage.
Hodgetts, T., and Lorimer, J., 2014. Methodologies for animals' geographies: cultures, communication and genomics. *Cultural Geographies*, 22 (2), 285–295.
Jasanoff, S., 2004. The idiom of co-production. *In*: S. Jasanoff, ed. *States of knowledge: the co-production of science and social order*. London: Routledge, 1–12.
Kirksey, S. E., and Helmreich, S., 2010. The emergence of multispecies ethnography. *Cultural Anthropology*, 25 (4), 545–576.
Latour, B., and Weibel, P., eds., 2005. *Making things public: atmospheres of democracy*. Cambridge, MA: MIT Press.
Law, J., 2004. *After method: the mess of social science research*. London: Routledge.
Lestel, D., Brunois, F., and Gaunet, F., 2006. Etho-ethnology and ethno-ethology. *Social Science Information*, 45 (2), 155–177.
Lury, C., and Wakeford, N., eds., 2013. *Inventive methods: the happening of the social*. London and New York: Routledge.
Moore, N., 2015. *The changing nature of eco/feminism: telling stories from Clayoquot Sound*. Vancouver: UBC Press.
Ogden, L. A., Hall, B., and Tanita, K., 2013. Animals, plants, people, and things: a review of multispecies ethnography. *Environment and Society: Advances in Research*, 4 (1), 5–24.
Ostrom, E., 1996. Crossing the great divide: coproduction, synergy, and development. *World Development*, 24 (6), 1073–1087.
Pryke, M., Rose, G., Whatmore, S., eds., 2003. *Using social theory: thinking through research*. London: Sage.
Plumwood, V., 2002. *Environmental culture: the ecological crisis of reason*. Abingdon, Oxon: Routledge.
Plumwood, V., 2007. Human exceptionalism and the limitations of animals: a review of Raimond Gaita's *The philosopher's dog*. *Australian Humanities Review* [online], 42 (August). Available from: http://www.australianhumanitiesreview.org/archive/Issue-August-2007/EcoHumanities/Plumwood.html [Accessed 25 May 2016].
Pollan, M., 2013. The intelligent plant. *The New Yorker*, December 23 & 30, 92–105.
Reason, P., 2005. Living as part of the whole: the implications of participation. *Journal of Curriculum and Pedagogy*, 2 (2), 35–41.
Roe, E., and Buser, M., 2016. Becoming ecological citizens: connecting people through performance art, food matter and practices. *Cultural Geographies*, online first, 1–16.
Savage, M., and Burrows, R., 2007. The coming crisis of empirical sociology. *Sociology*, 41 (5), 885–899.
Stengers, I., 2015. *In catastrophic times: resisting the coming barbarism*. Lüneburg: Open Humanities Press.
Tallbear, K., 2011. Why interspecies thinking needs Indigenous standpoints. *Cultural Anthropology* [online]. Available from: http://www.culanth.org/fieldsights/260-why-interspecies-thinking-needs-indigenous-standpoints [Accessed 25 May 2016].
Taylor, H., 2013. Connecting interdisciplinary dots: songbirds, 'white rats' and human exceptionalism. *Social Science Information*, 52 (2), 287–306.
Tsing, A. 2005. *Friction: an ethnography of global connection*. Princeton: Princeton University Press.
Wiles, R., Pain, H., and Crow, G., 2010. Innovation in qualitative research methods: a narrative review. *NCRM Working Paper Series*, 03/10.
Wolch, J. R., and Emel, J., eds., 1998. *Animal geographies: place, politics and identity in the nature-culture borderlands*. New York: Verso.

Part I

Experiments in more-than-human participatory research

1 Towards a more-than-human participatory research

Michelle Bastian

Introduction

In *Spindrift: A Wilderness Pilgrimage at Sea*, participatory action researcher Peter Reason (2014, p. 71) tells the story of a seminar offered by philosopher Henryk Skolimowski at the University of Bath. While supporting the participation of a wider range of people in research has been, and continues to be, a significant challenge to academic norms, this seminar offered further challenges by exploring participatory ways of engaging with the more-than-human world.[1] As Reason writes, 'this was strange, even to me. In the abstract world of university seminars, participation was still what one did with other people. It had nothing to do with the natural world' (2014, p. 71). While recognising the strangeness, a core question for this chapter is how participatory research (PR) might move in this kind of direction. Arguably issues like climate change, biodiversity loss and increasing rates of extinction create conditions where it is possible to put nonhumans explicitly on the PR agenda, and to ask how the commitments of PR – to situated knowledges, a wider recognition of agency and an expansive sense of stakeholders – might be revisited. That is, these crises invite participatory researchers to explore whether the injunctions of Western anthropocentrism might have unnecessarily restricted how participation is imagined, and to reconsider to whom its commitments might be made.

One way of supporting such enquiries is to bring PR and emerging more-than-human approaches into direct conversation. As noted in the introduction to this collection, both Henry Buller (2015), and Timothy Hodgetts and Jamie Lorimer (2015) argue that more-than-human geographies should seek methods that enable researchers to ask 'what matters' to nonhumans (e.g. Buller 2015, p. 7). For PR, this kind of question has been continually at its core, as participatory researchers seek ways of working with specific human communities to identify and respond to issues that matter to them. They do this by breaking down the boundaries between researcher and researched, ideally working in partnership to set research questions, determine which methods to use, analyse data, co-create outputs and develop dissemination strategies. In the process, broader questions of ethics, voice, knowledge and power are explored both practically and theoretically. Related questions also reside at the heart of more-than-human approaches, with issues of ethical relationality, the problem of representation, of exchange across different perceptual worlds and anthropocentrism constituting some of the area's most pressing issues.

These potentially fruitful overlaps between PR and more-than-human research (MtHR) were explored in a project called *In conversation with. . .: co-designing with more-than-human communities*, which took place in the UK in 2013, and which will be the focus of this chapter. Its two key objectives were first to ask whether participatory methods might extend towards a consideration of the more-than-human, and second whether the wealth of experience gained by participatory researchers, from working across social, cultural and other boundaries, might helpfully illuminate issues faced by more-than-human researchers. In order to respond to these questions we trialled the use of participatory methods, such as participatory design and participatory action research, as frameworks for two-day workshops with nonhumans. We wanted to know what might result from attempting to work with particular animals, insects, plants and elements specifically as *research partners*, rather than as subjects of experiments, for example.

This chapter will therefore share some of the insights generated by the project, as notes towards a more extended conversation about the possibility of more-than-human PR (MtH-PR). First, I outline the design and implementation of the project. I then place the project into conversation with PR literatures in order to highlight some of the ways that participatory approaches may indeed be open to working with wider understandings of who could be involved. Crucially, these literatures also offer cautions against the assumption that certain forms of inclusion are necessarily a good, and so this chapter will also discuss potential pitfalls of uncritically taking up the promise of participation.

Speculative field experiments

In looking for ways to describe the overall approach of the *In conversation with. . .* project, I would suggest that it might be thought of as a kind of philosophical field experiment (Bardini 2014, Frodeman *et al.* 2012), a form of speculative design (Dunne and Raby 2013) or perhaps even as a fantastic ethnography (Galloway 2013). That is, the project was not designed to establish MtH-PR as a definite possibility, since we were only at a preliminary exploratory stage. Instead, we were drawn to the speculative 'what if?' What if you could do participatory design with dogs? What if you could do participatory action research with bees? That is we primarily saw the workshops as putting ourselves in a position where we would be confronted with what it might mean to even *try* to include nonhumans in PR processes. In particular we were inspired by Clara Mancini and her colleagues, who argue that in seeking to conduct interspecies research there needs to be a willingness to explore the issues raised 'with genuine curiosity, no matter how challenging or ironic they may appear' (2012, p. 9). Thus even while recognising the stretching and cracking our questions might create within mainstream conceptions of what PR is and what it can do, we sought to take the tenets of both participation and the more-than-human as seriously as possible, put them into action, and see how this speculative experiment might play itself out.

Specifically, the project involved four exploratory workshops that took place between April and October 2013 in various locations in the UK. (Descriptions of each of these workshops can be seen in Boxes 1.1 and 1.2.) Attendees came from three main groups. First were members of a core team that included researchers from computing, environmental arts, forestry, geography, philosophy, sociology, theatre and women's studies, with further diversity in terms of the interdisciplinarity of their backgrounds and research methods used. Almost all were involved in the UK Arts and Humanities Research Council's Connected Communities programme, which has a particular focus on PR and which funded this project. Second, there were the nonhuman participants. In broad terms they included animals, insects, plants and the elements; more specifically dogs, bees, trees and water. The focus on these four was partly shaped by the expertise of the team and our preexisting links with potential partner organisations. However, we were also interested in pushing the boundaries of who, or what, could potentially be considered as an active research partner, and so the workshops focused on nonhumans across a range of (commonly assumed) levels of sentience, even though we also sought to trouble this hierarchy. Third were human intermediaries, such as dog trainers and beekeepers, who shared their expertise and facilitated engagements with the particular nonhumans that they worked with. Here we drew parallels between our project and the role of community leaders or community experts within PR, as well as more general discussions of border-crossers who are able to link different social worlds (Anzaldúa 1990).

As for the specific content of the workshops, we aimed from the outset to support diverse 'ways of knowing' (Graham *et al.* 2015), and so avoided the usual focus on academic presentations in favour of learning from the nonhuman participants and human intermediaries via inductions, practical/experiential activities and facilitated discussion and reflection. The workshops thus included at least one day of exploration, which was experiential and hands on. This included inspecting beehives, wild swimming and wood carving. These activities were analogous to the project initiation phase of PR where potential research partners spend time getting to know each other and exploring issues that are important to the community partners. Next, the core team and participating intermediaries articulated issues that arose during these activities and tried to identify which ones might develop into research questions, again drawing analogies with the later stages of project initiation.

We then workshopped a particular participatory model (see Box 1.1), keeping our commitment to our speculative approach always in mind. This often meant working through a specific PR handbook or toolkit and identifying what affordances or frictions might arise if groups tried to apply the guidelines in a project with a specific nonhuman partner. Some conversations that resulted included: the possibilities of data-gathering with bees, where we felt there might be some interesting approaches that could be developed; or asking whether core principles of participatory ethics, such as privacy, would hold when working with water, where we found it almost impossible to develop any kind of coherent response. Both of these kinds of responses were important as they helped to shape our understanding of how PR might extend towards a consideration of the more-than-human and

Box 1.1 The *In Conversation with...* workshops

1 *In conversation with animals* (April 2013) was organised by computer-interaction researcher Clara Mancini and philosopher Michelle Bastian and drew on a participatory design framework. It was conducted with the team from the Open University's Animal–Computer Interaction Lab and dogs and people from Dogs for Good (formerly Dogs for the Disabled). Activities included train the trainer exercises and interacting with service dogs and dogs in training.
2 *In conversation with insects* (May 2013) was organised by geographer Phil Jones and drew on a participatory action research framework (specifically, Pain et al. 2012). It was conducted with bees and people from the Evesham Beekeepers Association, as well as the Vale Heritage Landscape Trust. Activities included hive inspections and bee habitat maintenance.
3 *In conversation with plants* (September 2013) was organised by landscape and forestry researcher Richard Coles and drew on a community participatory arts perspective. It was conducted with the trees and people from the Wye Valley Area of Outstanding Natural Beauty and Wildwood Coppice Crafts. We explored techniques used by the Wye Valley InsideOUT project to connect excluded and under-represented groups with the forest. Activities included wood carving, materials collecting and making, as well as individual time spent in the woods.
4 *In conversation with the elements* (October 2013) was organised by geographer Owain Jones and artist Antony Lyons and used a set of ethical guidelines on community-based research developed within the Connected Communities programme (Banks and Manners 2012). It was conducted with water, specifically the River Torridge and its catchment area, and the people of the North Devon Biosphere Reserve, Devon Wildlife Trust and skipper Dave Gabe. Activities included field trips to the culm grasslands and a search for the river source, a boat trip up from the river's mouth, salinity sampling to see the mixing of fresh and sea water in the river, and wild swimming.

Detailed accounts of each of these workshops, including images and films, are available on the project website (www.morethanhumanresearch.com).

how it might not. They also highlighted which nonhumans might be more readily included than others and in what ways.

Finally, each workshop also included a session where we stepped back from the speculative experiment and critically reflected on the process. Here we explored the differences between the 'what if' and the 'what was'. While some of these

Box 1.2 A detailed look at *In Conversation with Dogs*

When starting to plan the workshops themselves, we found that turning the wider inspirations and approaches for the *In conversation with. . .* project into a programme of activities required its own kind of translation work. Faced with the task of designing the first workshop, both Clara and I found ourselves puzzling over what we were actually going to do.

Our plan was to build on work Clara had been doing with the Animal–Computer Interaction (ACI) Lab at the Open University (see Mancini, this volume). ACI arises out of the field of Human-Computer Interaction (HCI), which focuses on designing interactions with technology that are experienced positively and respond to the needs of specific user groups. However, it is rarely acknowledged that nonhuman animals can also be users of technology. For example, service dogs can learn to use kettles, washing machines and even cash machines. The ACI Lab thus seeks to design for specific nonhuman users by taking into account their physical and perceptual abilities, how they learn and what constitutes positive feedback for them.

We were also lucky to have keen and interested partners for the workshop. Namely, Dogs for Good, including service dogs Winnie and Cosmo, as well as Helen McCain, Head of Canine Training, and Duncan Edwards, Head of Client Liaisons, both of whom acted as our human mediators.

A core participatory approach within HCI is participatory design and so this was taken as our framework. Our building blocks were the core steps of the design development cycle, which include collaboratively identifying requirements, proposing designs, prototyping these designs and evaluation. As with the participatory methods we drew on for the other workshops, these steps require considerably more time than is available in a two-day workshop.

However within design there are also a range of methods that can create quick initial responses to a design problem. These include techniques such as paper prototyping, where initial design proposals are mocked up on paper, or design challenges where participants may cycle through the design process in a few hours to explore new ideas. Methods like these allow participants to get an initial sense of what kinds of tactics might work as well as what potential problems or blocks might arise. Given that our aim was principally to explore the potential for a dialogue between MtHR and PR, these kinds of approaches resonated well.

As became customary for each of our workshops, the teams were first sent a series of preparatory readings. These looked at issues of dog perception and evolution (Range et al. 2008, Honeycutt 2010, Taylor *et al.* 2011, van der Zee *et al.* 2012), examples of design focusing on dogs or human-dog interactions (Resner 2001, Mankoff et al. 2005, Wingrave *et al.* 2010, Higgin 2012) as well as texts on participatory design itself (Kensing and Blomberg 1998, Muller 2009).

The first day was spent 'identifying requirements.' This including presentations from Clara about HCI and ACI, as well as from Helen and Duncan who talked about the service relationship, how dogs learn their tasks, the different technologies they may interact with and some key issues that service dogs may face in their work. Helen and Duncan then led us in 'train the trainer' type activities where participants took turns being a 'trainer' or a 'dog' in order to practice clicker training, as well as attempting to navigate the workshop space using a wheelchair.

After finishing our induction, Winnie and Cosmo joined us in the workshop to work on 'problem definition'. This time was relatively brief since Helen and Duncan did not want to overtax the dogs by bringing them into an unfamiliar environment or involving them in activities for too long. Even so, we were able to include a session where the human and nonhuman participants were able to interact relatively freely, as well as the dogs working with the human participants to demonstrate some of the interactions that occur within a service relationship. Around mid-afternoon the Dogs for Good team wrapped up their contributions to the workshop and the remaining participants articulated problems encountered in these interactions, as well as broader issues, difficulties and complications as part of further refining the problem definition stage.

Day Two encompassed the remaining stages of the design development cycle, namely by proposing designs, prototyping these designs and evaluation. For this, Clara developed two design briefs in consultation with Helen and Duncan and arising from our problem definition work from the day before. Two challenges for service dogs with Dogs for Good include operating doors and buttons, since they can operate in a variety of ways and are not designed in a uniform way. We split into two teams and used the knowledge we had gained so far to propose a user-centred design solution to these challenges, including developing paper-prototypes. After feeding back to the wider group we then split up again to think through how we might work with the dogs from Dogs for Good to evaluate and refine the proposals. One suggestion was to offer a dog a series of options for operating doors and see whether one was selected more often, or could be used more quickly and easily.

Our last activity was to step back from the design experiment and reflect on the activities and processes overall, and particularly what insights our experiences might offer into the possibility (or not) of more-than-human participatory research.

reflections will be discussed below (and see also Heddon, in this volume), an example of an issue that arose was around the freedom of nonhumans to participate. Questions that came up included the following: Was inspecting a hive really analogous to meeting a community partner? What did it mean that we wore protective suits and used smoke to avoid being stung? Was wood carving a useful way to participate with trees and learn about their qualities, or was it more similar to a dissection? These questions were indicative of the generative nature of our discussions, and the impossibility of any quick and easy answers.

Analogies and diffractions: approaches to reading PR and MtH together

Having set out the project itself, and before analysing it in more depth, I want to briefly suggest some frames for understanding the kind of claims I will be making about potential inter-relationships between PR and MtHR. Drawing on my background in philosophy, one way this field experiment in MtH-PR can be read is as a particular form of analogical argument.[2] That is, it sheds light on both fields by exploring their similarities and differences with the other. As philosopher Paul Bartha notes, analogies can play 'an important heuristic role, as aids to discovery,' in part because they can be used 'to generate insight and to formulate possible solutions to problems' (2013, n.p.). In this sense an analogical reading of the two fields might open up unexpected proposals, such as drawing on PR to address more-than-human researchers' interests in asking 'what matters' to nonhumans. Further, Bartha suggests that analogies can also be important when proposing something that might at first appear strange, or even nonsensical, from common sense points of view. That is, 'often the point of an analogical argument is just to persuade people to take an idea seriously' (2013, n.p.): For example, taking seriously the idea that all those affected by research have a stake in the research, including nonhumans.

Within philosophy the key to a convincing argument-by-analogy is that there are enough parallels between the two cases to support the extrapolation of characteristics from one to the other. That is, a known similarity between the cases is used to extrapolate other points of similarity and sameness. One such similarity may be found in the respective commitments of MtH and PR approaches to including those traditionally excluded from research processes. Within traditional forms of analogous reasoning developing such points of connection encourages one to seek out further examples of congruence. This would suggest that recognition of an inter-relationship between these approaches rests on proving their similarities to each other.

However, a move towards proving similarity would arguably be antithetical to practices of working across difference and diversity in ways that are attentive to multiple and conflicting needs (e.g. Reagon 1983). Thus it is also important to look to contemporary reworkings of analogous reasoning which offer more felicitous approaches.[3] One particularly well-known reworking is Donna Haraway and Karen Barad's development of diffractive logics, which contrasts with an optics of reflection by enabling a shift away from a problematic emphasis on static identity, towards the processual effects each might have on the other (Haraway 1992, p. 300). Under this logic, the aim would not be to prove that MtH and PR approaches are sufficiently similar to each other to support exchanges of ideas between the two, but rather to ask whether the process of producing an 'interference pattern' between the two can create beneficial insights into 'how different differences get made, what gets excluded, and how those exclusions matter' (Barad 2007, p. 30). In this way, diffractive logics retain the benefits of analogies outlined by Bartha (i.e. as aids to discovery, generators of insight, persuasive supports) while avoiding the expectation of sameness. With this in mind I want to now turn to an analysis of the project itself.

Broadening participation

The first interference pattern I offer here is created by reading a recent review of PR with an eye towards the more-than-human. My focus is Jarg Bergold and Stefan Thomas' (2012) 'Participatory research methods: a methodological approach in motion', which provides an overview of the field and suggestions for development. Focusing on a review article enables me to engage with attitudes and approaches that are widely accepted and thus generally representative of the approach overall. This is important because I want to suggest that core features of the participatory approach, when viewed in light of the *In conversation with . . .* project, do indeed resonate with efforts to explicitly include nonhuman participants. In this way I want to suggest some initial ways that an explicit focus on the more-than-human (as research partner) might help to move participatory research in 'strange' directions.[4] The first two suggest ways that the entanglements of human participants in PR with nonhumans could be made more explicit, while the second two open up possibilities for an explicit engagement with nonhumans specifically.

Expanding life-worlds

In their review, Bergold and Thomas' initially define PR as being 'geared towards planning and conducting the research process *with* those people whose life-world and meaningful actions are under study' (2012, §1). While the focus is clearly on humans, the latter part of the definition, which emphasises meaningfulness and differences between life-worlds, suggests a shared epistemological approach with more-than-human research. This shared approach emphasises foregoing the search for universal truths and instead attending to specificities of experience and context. Indeed the reference to life-worlds calls to mind Jakob von Uexküll's (2010) notion of '*umwelt*' and his observation that 'we comfort ourselves all too easily with the illusion that the relations of another kind of subject to the things of its environment play out in the same space and time as the relations that link us to the things of our human environment' (2010, p. 54).[5] Such an observation resonates with PR's own critiques of hegemonic knowledge production and of the ability of objectivity and detachment in social scientific research to work with the diversity of human experience.

However, beyond this conceptual link, which draws an analogy between theoretical approaches, a diffractive view also points to what potentially gets excluded from the PR concern with people's life-worlds. In our project a key observation was the way that the life-worlds of our human intermediaries were not radically separated out from the nonhumans that they worked with. A good example of this was during the *In conversation with insects* workshop where the beekeepers suggested that working with bees had changed their behaviours, and their perceptions of the environment. For example, because of their concern for the bees' welfare, their life-worlds now included a greater awareness of the weather and the availability of forage. Thus, our project highlighted the way people can be 'differently human', as Niamh Moore (2013) put it, depending on how their lives are shaped by the various human and nonhuman agents that play a role in their life-worlds. When 'planning and conducting the research process' with the people under study, then, a MtH lens would challenge the exclusion of this more expansive field of stakeholders. Instead a diffracted PR might more explicitly recognise the nonhuman actors that also participate in and shape the life-worlds of the people in question.

Supporting cognitive estrangements

A further aspect of PR that Bergold and Thomas' outline is the ability of PR to create experiences of estrangement. This is important because it makes room for challenging embedded assumptions about how the world works, particularly assumptions held by those with more power. That is 'the participatory research process enables co-researchers to step back cognitively from familiar routines, forms of interaction, and power relationships in order to fundamentally question and rethink established interpretations of situations and strategies' (Bergold and Thomas 2012, §1). Here too we can make analogies with MtH research, which challenges fundamental assumptions of human exceptionalism and the 'forms of interaction' with the more-than-human world that it supports. Thus, when MtH and PR were brought into conversation in the context of the project, we found that the power relationships between humans and nonhumans could also be foregrounded and questioned.

This is illustrated first by a further example from the *In conversation with insects* workshop. In reflecting on the session one of the beekeeper participants, Martyn Cracknell, President of the Worcestershire Beekeepers Association, commented:

> I have been an amateur beekeeper for over 40 years, and I have always considered myself to be quite caring and empathic. I am very fond of my bees. When a colony requires management, e.g. to avoid losing a swarm, there are often several different strategies that might be used to achieve the desired end result. Ordinarily my choice would have been made by considering the efficacy of the method, the convenience for me, the timeframe for the operation

> and so on, but as a result of the workshop discussions I am now more mindful of how closely my intervention accords with the bees natural behaviour, and whether my intervention is sympathetic to the bee's needs. I had never really stopped to think about this before.
>
> (personal communication, 13 May 2016)

Second, in some cases, established cognitive frameworks were so thoroughly challenged that they were rendered almost absurd. Our efforts to think though the ethics of community-based research with water (i.e. Banks and Manners 2012, see also Banks *et al.* 2013), for example, which included discussing issues such as informed consent and anonymity with research partners (in our case, water), provoked as much silence as discussion. That is, the participatory framing pushed us so far away from familiar 'forms of interaction' that we found we had almost no conceptual frameworks to draw on.

While this may raise the criticism that we were trying to apply frameworks in contexts where they were simply not suited, this disorientation proved fruitful in that it allowed us to ask questions of the guidelines themselves. For example, having moved towards a position where our watery project partner was seen as inseparable from the other systems it is a part of (i.e. from an abstract 'water', to the specific Torridge watershed, supported by our shared reading of Linton [2010]), some participants wondered to what extent liberal notions of individual rationality might persist in the guidelines, which in our context were deeply problematic. This potentially opened up broader critiques of the way the subject is traditionally conceptualised within PR and illustrates a further way that PR might benefit from its interference pattern with MtH research.[6] Most importantly for us, however, this challenge could potentially be met by emphasising resources internal to PR, specifically the emphasis on cognitive estrangement.

The two preceding discussions of life-worlds and cognitive estrangements suggest potential ways that PR approaches can highlight the entanglements of human participants with more-than-human worlds. However, a further interest of the *In conversation with . . .* project was in the possibility of working with nonhumans directly. As such, the fundamental question remains of whether nonhumans could participate in research in ways that might be recognised as a form of PR. While this seemed highly doubtful during *In conversation with the elements*, other workshops suggested greater possibilities. This in itself points towards the need to augment our research questions further, and to attend to questions of which nonhumans are potentially involved, and what kinds of ways participation might be reconsidered for each of them. As moving from animals to insects to plants to the elements demonstrated for us, the question of what PR might offer to specific nonhuman actors needed to be asked again within each hoped for collaboration. More generally, then, finding responses to these questions requires exploring whether particular nonhumans have competencies that could support their involvement in PR, and whether PR could develop methods that would support any such competencies. Continuing with our diffractive reading of Bergold and Thomas's paper provides support for both explorative forays.

Challenging assumptions of competency

As Bergold and Thomas discuss, within PR there is a long history of rejecting claims that particular groups lack the competency for engaging in research. This includes challenging suggestions that they may have deficits in ability, or lack the appropriate social capital. Recognising that claims of a deficit are most often made by those in power, Bergold and Thomas suggest that from the perspective of PR 'the difference between the academic worldview and that of the research partners from the field is actually an asset which must be exploited in the exploration process' (2012, §42). Indeed they suggest, in reference to some of Bergold's earlier work, that 'participatory research can be regarded as a methodology that argues in favour of the possibility, the significance, and the usefulness of involving research partners in the knowledge-production process' (2012, §2). Within PR, then, competency is not a *fait accompli*, but an open and evolving question that further requires researchers themselves to reconsider their own competencies and develop capacities appropriate to the specific research context.

For the *In conversation with. . .* project, starting from an orientation towards possibility, rather than assuming from the outset that nonhuman participation was impossible, led to a number of insights into ways that different nonhumans might potentially contribute their worldviews to a research project. At *In conversation with animals*, for example, we explored the possibility that assistance dogs could provide feedback on prototypes designed to respond to issues they encounter in their work. Mancini and other members of the ACI Lab reminded us that there are already ways of working with pre-verbal or non-verbal humans that might provide useful insights, but also that situations could be designed that would suggest a dog's preference for one prototype over another. When discussing this at our *In conversation with insects* workshop, in a session on working with bees during the evaluation stage of a project, one of the beekeeper participants pointed out that assessing preferences for prototypes had also been adopted by biologist Thomas Seeley (2010) to try to understand which design of beehive particular hives might prefer. Understanding the significance and usefulness of MtH-PR (for the nonhuman as well as the human partners) may very well draw on the initial orientation towards competency-as-possibility that is at the core of PR, particularly in combination with the 'genuine curiosity' that Mancini and her colleagues have argued for.

Designing methods for inclusion

Building upon a basic curiosity in relation to competency requires the further step of developing methods and frameworks that are suitable for working with a research partner in light of their specific capabilities and needs. Such a statement would not be unfamiliar within PR circles, with Bergold and Thomas emphasising that PR places the onus on those designing a project to find ways for stakeholders to be involved, even if this means developing new approaches and techniques in order for them to do so. One example they discuss is mental health and disability

PR where concerns have been raised about the tendency to work with health professionals rather than with those directly affected by an illness or disability, in part because the latter may be 'in a very poor position to participate in participatory research projects, or to initiate such a project themselves' (Bergold and Thomas 2012, §20). However, self-advocacy groups of mental health service users have argued for the need to produce research that is independent of professional and institutional providers because of concerns around the hegemony of the medical model or entrenched economic interests within health-care (Bergold and Thomas 2012, §22). Thus the ideal of participation continues to push PR to develop ways of including those who have been rendered 'quasi-invisible' (Bergold and Thomas 2012, §26) and to innovate methodologically in order to do so. The *In conversation with. . .* project pushes such questions even further by highlighting the way that whole hosts of 'earth others' (Plumwood 1993, p. 156) have been rendered 'quasi-invisible' (if not just plain invisible) within PR. However reading PR diffractively also suggests that the ideal of doing research with those who are often unseen by dominant actors may very well support the experiments with method and approach that would be necessary to recognise the specific interests and needs of particular nonhumans.

The dangers of participation

The discussions above highlighted the ways that core ideals underpinning PR might be re-read as opening onto a much wider field of 'participants' than is usually supposed. Even so, while the ideal of participation emphasises inclusion and empowerment, critics of PR have pointed out that it can often be mobilised in highly programmatic and narrow ways. For example, Bill Cooke and Uma Kothari's (2001) collection *Participation: the new tyranny?*, has called attention to its problematic institutionalisation within the development context. Contributors to their collection question the presumption that participatory methods always unlock hierarchies and suggest that, in fact, they can maintain them. As Cooke and Kothari set out in their introduction, the aim of the collection is take a step back from the internal critique that is a core part of the participatory model itself, and instead ask fundamental questions about the approach as a whole (2001, pp. 1–2). Given the unclear use of terms such as participation, co-production and co-design within MtH research (see Introduction), the second interference pattern I want to set into motion draws on these types of critiques to focus on some of the potential dangers of using participation as a framework for working with nonhumans. That is, while the sections above used insights from the *In conversation with. . .* project to pose questions to PR about what is excluded from it, here debates within PR encourage a deeper questioning of the project itself.

Assuming participation is beneficial

In her contribution to Cooke and Kothari's collection, Frances Cleaver (2001) asks perhaps one of the most fundamental questions for PR, namely whether participation can be considered to be intrinsically beneficial. Discussing common understandings of the types of incentives for involvement in development projects, she notes that it is generally assumed that participation is in people's rational interest either because of the 'assurance of benefits to ensue' or because it is 'socially responsible and in the interests of community development as a whole' (Cleaver 2001, p. 48). Cleaver argues, however, that these assumptions are simplistic and more attention needs to be paid to the costs of participation and the benefits of refusing to participate. Indeed she points out that 'there are numerous documented examples of situations where individuals find it easier, more beneficial or habitually familiar not to participate' (Cleaver 2001, p. 51). Indeed within all human-based research there is a duty to support non-participation as an option. Participant information sheets often carry phrases like 'you are free to withdraw at any time without negative consequence' and contributions to a research project (participatory-based or not) are supposed to be free of any kind of coercion. That this is not always the case for humans, as Cleaver argues, suggests that attention to the option of non-participation should be even more important with nonhumans who are often in positions of significantly less power and autonomy.[7]

Arguably then, prior to focusing attention on 'what matters' to animals as Buller suggests, there should first be a consideration of the negative consequences such an investigation might have for the animals (and indeed other nonhumans) themselves. Indeed, as was noted at the time, within the *In conversation with. . .* project, smoke was used to pacify bees in order to inspect their hives, a cherry tree was cut down to provide wood for our wood carving activity and dogs had already been trained (and bred) to consent to the activities. As Clara Mancini suggested in our discussions (see also Mancini, this volume), the 'right to withdraw' could be understood as one of the key dividing lines between collaborative knowledge seeking and animal experimentation. In what way these considerations might apply to plants or elemental partners remained an open question. This suggests that any MtH-PR would need to ask to what extent participation is simply being assumed to be a 'good thing' and to interrogate the initial impulse toward 'inclusion' further.

Overlooking wider inequalities

A further critique posed by the Cooke and Kothari collection centres on the scope of participatory projects done with marginalised communities, and particularly the narrow focus of many development projects. The worry is that 'an emphasis on the micro level of intervention can obscure, and indeed sustain, broader macro-level inequalities and injustice' (Cooke and Kothari 2001, p. 14). If PR is to challenge entrenched power structures then it cannot focus on the smallest or easiest interventions. This kind of claim resonated with some of our own questions in the project. *In conversation with animals* for example, focused on involving service dogs in co-designing tools that made their work easier, but had less room for exploring the service relationship itself. Our experience of 'meeting the bees' during the hive inspections was also based on a prior relationship between bees and beekeepers that some (though not all) might argue is fundamentally problematic. An important question for our project then was that in focusing on issues that seemed manageable (in part because what we were doing felt so unconventional) did we close off the option of tackling macro issues to do with the very nature of the relationship between the humans and nonhumans at the focus of our workshops?

Overall, I would suggest that there was an awareness of these kinds of broader issues, in part because of the wide variety of participants and the emphasis on physical interaction and critical reflection. *In conversation with insects*, for example, ended with discussions of the power structures within beekeeping associations and suggestions for the development of forms of 'co-responsible beekeeping' within them. However, given that ours was an explorative project, where we were speculating about tangible projects, a more concerted effort at a potential MtH-PR may very well encounter different pressures around what seems reasonable to tackle and what does not. Here then it would seem useful to draw on work in PR that explores how groups can tackle both the macro and the micro, as Virginia Eubanks (2009, p. 113) discusses in her work on popular technology.

Pseudo-participations

Earlier I suggested that the emphasis within PR on approaching competency in an open way, and innovating methods to support varying needs and interests seemed promising for a potential MtH-PR. However, by returning now to Bergold and Thomas' paper we can also see that in practice there has been a disparity between these ideals and who PR is most often done with. That is, they argue that PR is more common with and amongst professional practitioners, such as with mental health professionals, than with 'the immediately affected persons', such as mental health service users (Bergold and Thomas 2012, §19). This is in part because, despite aspirations for inclusion, the competencies of practitioners are still more likely to support their participation in, or initiation of, PR projects.

An awareness of this dilemma seems particularly important for a MtH-PR since it also appeared in the *In conversation with. . .* project. As some readers might have already noted, for all the project workshops the human mediators participated more fully over the two days, than did the dogs, bees, forest and river. Indeed for all of the workshops, our nonhuman partners were involved in the initial information gathering phase only, and the later stages of data-gathering, evaluation, dissemination and so on were completed by trying to imagine or extrapolate what might happen if researchers tried to support the inclusion of nonhumans in them.

In this speculative attempt at imagining what MtH-PR might mean in practice we focused on what Bergold and Thomas refer to as 'so-called "early" forms of participation, such as the briefing of professional researchers by those who are affected by the problem under study' (2012, §32). This could possibly be thought of as a step in the right direction, particularly if understood as 'preparatory joint activities that may facilitate participation in the research project at a later date'

(2012, §32). However, it should also not be discounted that what took place could alternatively be understood as 'pseudo participation' mobilised for ends other than those that seek to benefit the participant (2012, §32). Indeed in the later sections of the workshops many human participants commented on the difficulty of retaining a sense of what, for example, a bee's perspective might be in relation to the issues at hand. While further attempts at MtH-PR might still retain certain divisions of labour (again see Eubanks 2009, p. 115), it would seem important to identify any tendency to avoid the hard task of working out the possibilities of working with specific nonhuman partners or overly relying on human mediators.

No wider context of support

My fourth and final point is also suggested in Bergold and Thomas's review and raises the question of what kind of context might be needed in order to support these kinds of participatory experiments. At the outset of their review they argue that 'every type of research calls for social conditions that are conducive to the topic and to the epistemological approach in question' (Bergold and Thomas 2012, §32). They use this assumption to suggest that only within a broader political context of democracy is participation a viable research method. While these claims are made rather quickly in their article and would need to be explored more fully, the issue they raise is important for thinking about what broader social and political contexts might be necessary for MtH-PR to both appear legitimate and be viable. As suggested above, we had already found that some questions are easier to ask than others (e.g. the micro rather than the macro).

Further, the beliefs that nonhumans could (or indeed should) be treated as knowledgeable agents in their own right, and that they might have a stake in broader knowledge making processes, not only challenge political and social contexts but also many of the fundamental assumptions of Western ontological and epistemological frameworks, including the assumptions of many PR approaches. Indeed, as suggested previously, in the *In conversation with the elements* workshop we often felt that we ended up in a context where it was impossible to find frameworks or terminologies that could orient us when seeking to answer the strange questions put to us when drawing on PR approaches to work with water. These discussions in particular highlighted the liberal frameworks that many PR approaches draw on (e.g. justice, rights and inclusion being predicated on individual autonomy, agency and shared rational dialogue) and the ways that a potential MtH-PR would provoke questions around the humanism underlying them. As my colleague Franklin Ginn asked me, could the very idea of seeking to include nonhumans in PR not itself risk becoming what Cary Wolfe (2010) has called a 'humanist posthumanism', where their inclusion is judged according to humanist ideals? This again suggests that an MtH-PR would need to tackle macro issues, such as those of epistemology and ontology, at the same time as meso issues of methods and approaches, and micro issues of developing specific interventions in specific contexts.

Conclusion

This chapter has highlighted the potential of a diffractive reading of PR and MtHR for reconsidering the methodological possibilities of each. Adopting elements of the ethos underlying PR, such as insisting on the responsibility of researchers to ask which stakeholders are being excluded from the process, and on the non-neutral character of determinations of competency, could add a certain kind of boldness to MtHR. Further, internal and external critiques of PR could help MtHR think through the fraught nature of participation and the gap that can exist between theory and practice. While for PR, MtHR could push these approaches to question their focus on the human, and also to explore the differences between the liberal frameworks common to PR and frameworks of mutual entanglement more common within MtHR. Moreover, issues such as inclusion and exclusion, the contextual nature of knowledge, and the relationship between power, voice and agency have been central to both PR and MtH approaches and yet have been approached in very different ways. The *In conversation with. . .* project suggests that these interconnections are worthy of exploration, and hopefully future research will delve further into the strange patterns produced by the diffraction of each with the other.

Acknowledgements

Thanks to the whole project team for a brilliant year of explorations, challenges and laughter. Thanks to my friends and colleagues who have commented on this paper so carefully, including Niamh Moore, Emma Roe and members of the Edinburgh Environmental Humanities Network writing group. Thanks also to attendees at seminars where parts of this has been presented, including at the University of Melbourne, the University of New South Wales, two RGS-IBG conferences and at the University of Edinburgh. Finally special thanks to the AHRC's Connected Community programme for supporting *In conversation with. . .* (AH/K006517/1) and allowing us to take risks with our research.

Notes

1 See Reason, in this volume, for further details.
2 Noting that other project participants have used different frames (see, for example, Goto Collins and Collins, Heddon, Pigott and Lyons in this volume).
3 Thank you to Affrica Taylor for making this point and reminding me of the importance of diffraction in this context.
4 Here I want to note that these 'strange' directions might also draw on heritages within PR, such as research focused on sustainability and the environment, which, while not explicitly including the more-than-human as participant, often have the aim of making a positive difference in these more-than-human worlds. Further, participatory work with indigenous peoples has also emphasised the participation of the more-than-human (see Coombes *et al.* 2014, pp. 849–851). Thanks to Niamh Moore for prompting me to think about this more explicitly.
5 Importantly this should not be read as suggesting we therefore occupy radically separate spaces and times. That is, even while Uexküll likens the *umwelt* to a soap bubble

this is still in a context where 'relations between things expand and mesh with one another in intricate webs of life' (Buchanan 2008, p. 25).

6 During our discussions it was also recognised that the practice of PR often enacts more complex and fluid understandings of the subject (e.g. Eubanks 2009), and of processes such as consent (e.g. Dewing 2007). However, some also commented on a seeming disconnect between this and the subject that was assumed in the handbooks, toolkits and guidelines that we drew on for the project.

7 Although as we discussed in the *In conversation with the elements* workshop, nonhumans can also be much harder to coerce (see Bastian 2013). For example, those working with water ignore its capacities at their peril.

References

Anzaldúa, G., 1990. Bridge, drawbridge, sandbar or island: lesbians-of-colour *hacienda slianzas*. *In:* L. Albrecht and R.M. Brewer, eds. *Bridges of power: women's multicultural alliances*. Philadelphia, PA: New Society Publishers, 216–231.

Banks, S., and Manners, P., 2012. *Community-based participatory research: a guide to ethical principles and practice*. Bristol: National Coordinating Centre for Public Engagement.

Banks, S., Armstrong, A., Carter, K., Graham, H., Hayward, P., Henry, A., Holland, T., Holmes, C., Lee, A., McNulty, A., Moore, N., Nayling, N., Stokoe, A., Strachan, A., 2013. Everyday ethics in community-based participatory research. *Contemporary Social Science*, 8 (3), 263–277.

Barad, K., 2007. *Meeting the universe halfway: quantum physics and the entanglement of matter and meaning*. Durham & London: Duke University Press.

Bardini, T., 2014. Preface: a field philosopher with a certain taste for fish, and who does not mistake his hat for an ethology. *Angelaki*, 19 (3), 5–9.

Bartha, P., 2013. Analogy and analogical reasoning [online]. *In*: E.N. Zalta, ed. *The Stanford encyclopedia of philosophy*. Stanford, CA: Stanford University. Available from: http://plato.stanford.edu/archives/fall2013/entries/reasoning-analogy/ [Accessed 23 May 2016].

Bastian, M., 2013. *Co-designing with water* [online]. Available from: http://www.morethanhumanresearch.com/home/co-designing-with-water [Accessed 23 May 2016].

Bergold, J., and Thomas, S., 2012. Participatory research methods: a methodological approach in motion. *Forum: Qualitative Social Research*, 13 (1), Art. 30.

Buchanan, B., 2008. *Onto-ethologies: the animal environments of Uexküll, Heidegger, Merleau-Ponty, and Deleuze*. Albany, NY: State University of New York Press.

Buller, H., 2015. Animal geographies II: methods. *Progress in Human Geography*, 39 (3), 374–384.

Cleaver, F., 2001. Institutions, agency and the limitations of participatory approaches to development. *In:* B. Cooke and U. Kothari, eds. *Participation: the new tyranny?* London and New York: Zed Books, 36–55.

Cooke, B., and Kothari, U., eds., 2001. *Participation: the new tyranny?* London, New York: Zed.

Coombes, B., Johnson, J.T., and Howitt, R., 2014. Indigenous geographies III: methodological innovation and the unsettling of participatory research. *Progress in Human Geography*, 38 (6), 845–854.

Dewing, J., 2007. Participatory research: a method for process consent with persons who have dementia. *Dementia*, 6 (1), 11–25.

Dunne, A., and Raby, F., 2013. *Speculative everything: design, fiction and social dreaming*. Cambridge, MA and London, UK: MIT Press.

Eubanks, V., 2009. Double-bound: putting the power back into participatory research. *Frontiers: A Journal of Women Studies*, 30 (1), 107–137.

Galloway, A., 2013. Towards fantastic ethnography and speculative design. *Ethnography Matters* [online]. Available from: http://ethnographymatters.net/blog/2013/09/17/towards-fantastic-ethnography-and-speculative-design/ [Accessed 17 February 2015].

Graham, H., Hill, K., Holland, T., Pool, S., 2015. When the workshop is working: the role of artists in collaborative research with young people and communities. *Qualitative Research Journal*, 15 (4), 404–415.

Haraway, D., 1992. The promises of monsters: a regenerative politics for inappropriate/d others. *In:* L. Grossberg, C. Nelson, and P. Treichler, eds. *Cultural studies.* New York and London: Routledge, 295–337.

Higgin, M., 2012. Being guided by dogs. *In:* L. Birke and J. Hockenhull, eds. *Crossing boundaries: investigating human-animal relationships.* Leiden: Brill, 73–90.

Hodgetts, T., and Lorimer, J., 2015. Methodologies for animals' geographies: cultures, communication and genomics. *Cultural Geographies*, 22 (2), 285–295.

Honeycutt, R., 2010. Unraveling the mysteries of dog evolution. *BMC Biology*, 8 (20), 1–5.

Kensing, F., and Blomberg, J., 1998. Participatory design: issues and concerns. *Computer Supported Cooperative Work (CSCW)*, 7 (3–4), 167–185.

Linton, J., 2010. *What is water? the history of a modern abstraction.* Vancouver and Toronto: UBC Press.

Mancini, C., van der Linden, J., Bryan, J., Stuart, A., 2012. Exploring interspecies sense-making: dog tracking semiotics and multispecies ethnography. *In: Proceedings of the 2012 ACM Conference on Ubiquitous Computing 5–8 September Pittsburgh, PA*. New York: ACM, 143–152.

Mankoff, D., Dey, A., Mankoff, J., Mankoff, K., 2005. Supporting interspecies social awareness: using peripheral displays for distributed pack awareness. *In*: *Proceedings of the 18th annual ACM symposium on User interface software and technology 23–26 October, Seattle, WA*. New York: ACM, 253–258.

Moore, N., 2013. Geneaologies, promises/dangers and modest interventions. *Presentation at In Conversation with . . . reflective workshop.* Birmingham.

Muller, M.J., 2009. Participatory design: the third space in HCI. *In:* A. Sears and J.A. Jacko, eds. *Human-computer interaction: development process.* Boca Raton, FL: CRC Press, 165–186.

Pain, R., Whitman, G., Milledge, D., Lune Rivers Trust, 2012. *Participatory action research toolkit: an introduction to using PAR as an approach to learning, research and action.* Durham: Durham University.

Plumwood, V., 1993. *Feminism and the mastery of nature.* London and New York: Routledge.

Range, F., Aust, U., Steurer, M., Huber, L., 2008. Visual categorization of natural stimuli by domestic dogs. *Animal Cognition*, 11 (2), 339–347.

Reagon, B.J., 1983. Coalition politics: turning the century. *In:* B. Smith, ed. *Home girls: a black feminist anthology.* New York: Kitchen Table: Women of Color Press, 356–368.

Reason, P., 2014. *Spindrift: a wilderness pilgrimage at sea.* Bristol: Vala.

Resner, B.I., 2001. *Rover@Home: computer mediated remote interaction between humans and dogs.* Thesis (MSc). Massachusetts Institute of Technology.

Seeley, T.D., 2010. *Honeybee democracy.* Princeton, NJ: Princeton University Press.

Taylor, A.M., Reby, D., and McComb, K., 2011. Cross modal perception of body size in domestic dogs (*Canis familiaris*). *PLoS ONE*, 6 (2), e17069.

van der Zee, E., Zulch, H., and Mills, D., 2012. Word generalization by a dog (*Canis familiaris*): is shape important? *PLoS ONE*, 7 (11), e49382.

von Uexküll, J., 2010. *A foray into the worlds of animals and humans.* Minneapolis and London: University of Minnesota Press.

Wingrave, C.A., Rose, J., Langston, T., LaViola, Jr., J.J., 2010. Early explorations of CAT: canine amusement and training. *In: CHI'10 Extended Abstracts on Human Factors in Computing Systems, 10–15 April, Atlanta, GA*. New York: ACM, 2661–2669.

Wolfe, C., 2010. *What is posthumanism?* Minneapolis, MN: University of Minnesota Press.

2 Marginalized voices

Zoömusicology through a participatory lens

Hollis Taylor

Despite accelerating scholarly interest in the more-than-human, many disciplines continue to theorize their pursuits with an assumption of human exceptionalism. Musicology is a case in point, with its borders fully closed to animals. Although some composers appropriate animals' acoustic constructs, the focus is typically on the 'improvement' that these human composers supply. Given this disciplinary history, an appeal to acknowledge birdsong as music might be understood as political as much as aesthetic.

In this chapter, I note how the marginalization of avian voices (as musical instruments and as embodiments of individuals on a quest to survive and thrive) expands to include voices of some humans and technologies well positioned to report on songbirds but nonetheless routinely sidelined.[1] (The vast majority of birdsong researchers hail from the natural sciences and conduct studies in controlled laboratory settings.) In addition, I review how *zoömusicology* (the study of music in animal culture) might assist us to re-imagine and revise binaries like human/nonhuman, nature/culture, expert/non-expert, subject/object, and insider/outsider.[2] Similar dualistic constructions routinely confront participatory knowledge production processes, which I place in dialogue with zoömusicology in a search for correspondences or sparks of disparity – or both.

Participatory Research (PR) serves as an overarching rubric for diverse methodologies and ideals that are responsive to and co-shaped (or coproduced) by stakeholders. However, since both zoömusicology and PR reject universal designs with a one-size-fits-all approach, and are instead pluralistic in orientation and methodology, my task below is to move us from the general to the specific – to ways of knowing found in local settings and local individuals (Townsend 2013). In the process, I draw on approaches from participatory design, participatory geography, participatory ethnography, and elsewhere.

Music and *participation* are concepts that have been linked since time immemorial. In writing about the value of hands-on music-making, for instance, Anthony Everitt abandons terms like 'community' and 'amateur', which are often used to describe music created by or with 'non-experts', in favor of 'participation' and 'participatory' (1997, p.16). However, musicking is a negotiation, one made within a social structure. Calling music a 'participatory art' does not guarantee that it operates under the banner of PR. In fact, Everitt's self-described non-elitist approach must grapple with the nature of much contemporary music-making, which is essentially top-down. The master teacher, the bandleader, the conductor, the manager, the concertmaster/mistress, the tradition of a genre, the wealthy and powerful music industry, music competitions, unwritten rules, and issues of gender, age, and self-esteem – 'music participation' usually means being part of a group with a more or less rigid hierarchy.

Thus, in a search for contexts where music is *coproduced*, with musicians as facilitators, and where the aim of the project is to work together to develop a piece, we must visit what some consider the fringes of music-making. Jazz (and other improvised music) and music therapy are two notable arenas for coproduction. In fact, music therapist Kenneth Aigen has identified strong parallels between nonclinical improvised music (jazz in particular) and music therapy, namely 'the emphasis on process over product; the value placed on improvisation, spontaneity, creativity, intuition, and in-the-moment responsiveness; and, clearly differentiated and defined musical roles for participants that require a balance of freedom and structure' (2012, p. 180).

Such goals and methodologies see a natural overlap with PR, but many music therapists go further, explicitly adopting PR praxis. A growing trend would conceive of music therapy as social action.[3] For instance, Randi Rolvsjord has tracked the shift from an expert music therapist intervention to client-therapist collaboration (2006, p. 11). Participatory philosophies undergird Community Music Therapy (CoMT), which aims to offer a meaningful encounter that empowers society's 'others': the ill, the disabled, the aged, and the socially marginalized.[4] Some of the ways that PR variously reframes these 'others' is as *clients*, *users*, *end-users*, *community users*, *participants*, *stakeholders*, *fellow researchers*, *key partners*, and *design partners*.

Might we speak of a songbird as a *key partner* or *design partner*? As in music therapy, when PR methodologies and ideals are placed in dialogue with zoömusicology, the conversation is a lively and productive one. I offer up my ruminations in the form of a playbill.

Marginalized voices

A play in three acts.

The playhouse

Where we stage our production is key to how our play will unfold, who our audience will be, and how our critics will respond. Also crucial is whether we choose to coproduce our play or instead maintain sole control: Should birdsong study be exclusive to scientists working on laboratory-housed birds – or can we make the case for a zoömusicologist to conduct research with free-living individuals?

Science's classic laboratory playhouse

The scientific attraction of birds is that they are easier to study than many species. Intelligent, relatively low maintenance, and inexpensive to house and feed, birds also typically breed easily in captivity. Technological advances have facilitated birdsong research by zoölogists without complementary input from musicologists. In eschewing either/or propositions, I am not concerned with characterizing science as good or bad – clearly, significant work has been accomplished. Unfortunately, due to premature specialization by zoologists, most of what we believe we know about songbirds derives from a handful of smaller species like the zebra finch (*Taeniopygia guttata*), common chaffinch (*Fringilla coelebs*), white-crowned sparrow (*Zonotrichia leucophrys*), and domestic canary (*Serinus canaria domestica*). Considerable pressure exists to study them in a classic laboratory setting, where 'asking' an animal is based on the premise of confinement. However, such studies have begun to be problematized as merely suggesting what an individual might do in an environment unprecedented in that species' evolutionary history (Beecher 1996, p. 61; Kroodsman 1996, p. 4). In addition, when animals are routinely available, a researcher's position relative to their *object* changes – there is scant chance for an animal to be a *subject* under such conditions.

Zoömusicology's playhouse

For four months each year, I record the vocalizations and behaviors of pied butcherbirds (*Cracticus nigrogularis*) in outback Australia. Zoömusicology's outdoor playhouse finds *me* vulnerable (as well as responsible) to my research partners. Pied butcherbird spring nocturnal singing often necessitates sleeping in my car on site in order to commence recording in the hours before sunrise. Fieldwork knowledge is bodily knowledge, poles apart from the abstract knowledge of a laboratory researcher, who can forget she has a body. While even zoölogists who *do* conduct fieldwork must nonetheless adhere to standard models of scientific 'objectivity', as a zoömusicologist I can instead avow 'the partial and partisan' nature of my longitudinal enquiry (Merrifield 1995, p. 50).[5] Being with songbirds in the field (and acknowledging my presence and biography, also valorized in PR but systematically expunged in science's disembodied, detached reports) has been indispensable to my apprenticeship.

In matters of musicality, zebra finches and other current 'lab rats' are not interchangeable with free-living, sophisticated songsters – individuals likely unsuitable for laboratory housing.[6] Since a wealth of precision recording and measuring devices can accompany researchers into the field, I am content to give up laboratory supervision in the confidence that if my research depended on birds singing in such a setting, my understanding of their vocalizations would be vastly diminished. Minimizing distance, increasing empathy, real-world contexts – these are recurring themes of both zoömusicology and PR (e.g. Smith and Kjærsgaard 2014).

The actors

This casting of 'actors' is inspired by Bruno Latour's multi-realist Actor-Network-Theory (2005).

A zoömusicologist trying to find her way

As a zoömusicologist, I unapologetically embrace musicological tools and insights (including a trained ear and a musician's hunch), honor knowing but neglected voices from multiple species, and allow for rarities, one-offs, and anomalies inaccessible (or ignored) in the laboratory to play a part in investigations of animal capacities.

Technologies to record, document, and analyze birdsong

The sonogram has featured in birdsong studies since 1950; this graphic representation of sound has been championed for its nearly unquestioned, value-free objectivity. Like Latour's *immutable mobile* (an instrument that is easily readable, reproducible, translatable, combinable, and transportable), the sonogram 'mustered better and more data "allies" for the sciences than did the musical score' (Mundy 2009, p. 220). Listening (and transcribing by ear) slipped into a questionable category – one that also condemns the subjective, the feminine, the other, and the animal (in these, we are never far from PR's 'stakeholders'). As a result, it has become frowned upon to place birdsong in *conventional music notation*. However, this essential tool in my work allows me to bring birdsong into 'a sphere of discourse that is enabled by a distinguished intellectual history and undeniable institutional power' (Agawu 1995, pp. 392–393). When linked with technologies like sonographic analysis and other software systems designed to manage the acquisition and analysis of large amounts of data, this devalorized but powerful 'other' technology (along with musicology's analytic apparatus) has significant contributions to make in birdsong studies. Notation is *my* immutable mobile (Latour 1990).

A number of individual pied butcherbirds

While pied butcherbirds are celebrated as one of Australia's most accomplished songbirds, this is strictly anecdotal, with no in-depth studies of their vocalizations conducted prior to my research. (My point intends no disrespect to lay listeners who have commended the pied butcherbird voice; it is rather to underline that when theorists hold forth on songbirds' capacities, members of this species – and the vast majority of others – have not been formally canvassed.) Both sexes sing, including in duos, trios, and even larger ensembles. In the spring, soloists vocalize nocturnally for up to seven hours, and they transform these songs annually. Each bird is unique and not merely a kind; likewise, there is no single song – there is one song after another delivered by one individual after another (Taylor 2013). Local birds, local outcomes.[7]

Musicking

Human music-making (and not just clinical music therapy) has been presumed to produce emotional, physical, cultural, and social improvements even before metrics were put in place to measure them (e.g. Hillman 2002). *All* music is therapy. In contemplating this transformational potential, lawyer Joshua Dankoff brings a problematized music (no matter the genre) front and center into development discourse – one never straightforwardly instrumental or rational but with the potential for a political agenda. Although he reasons that 'music has been both a means of emancipation from domination but also instrumentalized as a tool for domination', he nonetheless urges that development be reconceptualized to include non-free-market experiences like music, rather than being preoccupied with the market and the rational individual (Dankoff 2011, p. 257). Disagreement about the nature and use of music abounds, and although any investigation into music requires input from diverse sources, musicology has kept a tight rein on its subject. In light of this, many (like philosophers Adrian Currie and Anton Killin [2015]) argue for concepts of music to be multiple and non-equivalent.

Human exceptionalism

Scholars increasingly urge a reconfiguration of institutional and other boundaries without the impediment of our 'problematic solace of human exceptionalism' (Haraway 2008, p. 46). The conventional view that only humans make music implies that only humans inhabit worlds of meaning and thrive by means of culture. Questions of who knows (and even of who pretends not – or chooses not – to know) are particularly germane: a multi-faceted understanding of music and musicality expands not only our understanding of what it is to be a songbird, but also our understanding of the human. It would seem perverse to deny the musical contributions of songbirds. In the midst of unstudied species, uncertain definitions, and incomplete statistics, there are nevertheless those who rule out ever being surprised by an animal (Lestel 2007b, p. 15).

Anecdote and anthropomorphism

Comments that speak to the musicality of a bird's song may be tagged 'anthropomorphic'. Sociologist Eileen Crist, however, argues that the 'singularly privileged perspective' of the scientific method cannot help but eliminate certain dimensions and thus distort the realities of an animal's life (2006, p. 206). Compelling narratives from outside a scientist's methodology or hypothesis-testing are 'officially unusable' (Dennett 1987, p. 250). On the other hand, a zoömusicologist's task embraces being open to a variety of narratives in order to form the most enlightened estimation possible (Taylor 2011). Likewise, participatory praxis does not rest on the reproducibility of findings, or on a neutral and privileged viewpoint; instead, empowerment of a local contingency depends upon honoring a range of unofficial stories.

The nature/culture crisis

Despite Copernicus and Galileo removing us from the center of the universe centuries ago, Descartes' consolidation in the Enlightenment of the nature/culture divide remains tightly, but almost invisibly, woven into the underpinnings of Western thought and discourse. Binary arrangements like nature/culture promote separation over continuity, prompting scholars to suggest that environmental fragmentation could be the result of our fragmented notion of the world. As ecofeminist philosopher Val Plumwood stresses, 'insights of continuity and kinship with other life forms (the real scandal of Darwin's thought) remain only superficially absorbed in the dominant culture, even by scientists' (Plumwood 2009, p. 120). In response, the fresh questions and different interpretive, analytical, and communicative methods of some humanities scholars exhibit strong parallels with PR. Author and social activist Naomi Klein describes her vision of climate change action like this: 'Fundamentally, the task is to articulate not just an alternative set of policy proposals but an alternative worldview to rival the one at the heart of the ecological crisis – embedded in interdependence rather than hyper-individualism, reciprocity rather than dominance, and cooperation rather than hierarchy' (2014, p. 462). Similarly, participatory geographers Hilary Gibson-Wood and Sarah Wakefield believe that 'in order to promote a just environmentalism, we must consider whose experiences and ways of knowing about the environment are included or excluded from our very definitions of "the environment" and "environmental problems" ' (2013, p. 645).

Enter birdsong, a biomarker (or indicator of a biological state) that reveals items of concern and consequence to birds and us. Birds' presence is a significant tool for measuring biodiversity, since a vocalizing bird indicates that the individual is both present and in sufficient condition to devote time and energy to activities not vital to quotidian survival. Both more-than-human and ethical human lives depend on us attending to marginalized voices so we may better grasp who we are and who we must become to rectify current ecological challenges. This attending goes right to the heart of PR: consulting a wide range of stakeholders helps us get our bearings (Gibson-Wood and Wakefield 2013), while failure to see ourselves as part of ecosystems places our 'basic survival project' in grave risk (Plumwood 2009, p. 128).

Having reviewed this pre-show bundling of people, animals, technologies, concepts, and values, we are prepared for our play to begin.

One: the setup, or coproducing methods

Synopsis

A zoömusicologist looks for ways of asking pied butcherbirds about their vocalizations.

Notes

'You need a good anatomist', a zoölogist scolded me after I presented a lecture on pied butcherbirds' vocal achievements to his research students. I had failed to peer inside my soloists' brains. At this key juncture, I determined that zoömusicological research could be rigorous while nonetheless stopping short of reacting to science's agenda. I benefit from some of science's knowledge claims and methods but remain wary of others. Explanations of hardware that require 'harvesting' a bird-musician's brain or placing them in a cage are, for me, a no-go zone. Can we speak of my ethical commitment to songbirds as *coproduction* – or does this concept need to be reworked in more-than-human research?

In concept and practice, PR is above all about the sharing of power (and its potential abuse), yet it cannot be summarily divested.[8] The thorny topic of power has caused much ink to be spilt by participatory researchers. Feminist STS scholar and activist Virginia Eubanks sees a participatory orientation as 'a starting point, an epistemological and ethical commitment, not a program or roadmap for practice' (2009, p. 115). Pertinent to my project here, participatory geographer Mike Kesby proposes that we 'draw on technologies such as participation in order to outmaneuver more domineering forms of power' (2005, p. 2038). Yet, human geographer Giles Mohan urges participatory researchers not to dismiss their core competencies in attempts to distribute power: 'by valorizing the local and being self-critical of our colonizing knowledge "we" behave as if we do not have anything to offer' (2001, p. 162). Whether in developing an animal study or a participatory design workshop, the ideal of an even playing field is elusive, as is a value-free environment.

Like participatory practitioners, it took me time to figure how to invent new methods or how to apply old methods in novel ways (e.g. Parfitt 2004). 'Our readers are not interested in creatures', a peer reviewer gave as the reason for rejecting my article for a music journal. Meanwhile, a science journal editor critiqued, 'Your work should be framed within a hypothesis-driven context'. Participatory praxis has also been confronted with instances where findings based on alternative methodologies and unofficial voices are blocked from a discipline's canon (e.g. Bergold and Thomas 2012).

For me as a zoömusicologist, how something is done must take priority over what is done. Likewise, Ole Sejer Iversen et al. argue that values (of users, stakeholders, and designers), while not fixed, must be the focal concern and thus influence the choice of methodologies (2012). I select methods and technologies that are best placed to highlight songbirds' achievements. Crucially, the song of each bird arrives with its own suggestions as to how it might be fruitfully analyzed.

Although I read scientists' birdsong reports, I never discount personal knowledge. The musician in me follows the concept of *kinship* wherever it takes me. My critical positions of both insider (a fellow musician) and outsider (from another species) jockey for position – a dance well-known to participatory facilitators – but when it comes to making music, how far apart do our two species dwell and how much mediation is required? Sometimes an intruder, sometimes an eavesdropper, sometimes a merely tolerated attendee at their pageant, I nevertheless believe that my physiology and sense of musicality overlap enough with pied butcherbirds to allow me entrance into their musical lives.

Entr'acte

A bird vocalizes at an open door while Steve sings and plays guitar – he entitled his home movie 'Butcherbird auditioning to join the band':

> On the video of the bird demanding breakfast, you will hear the very worst rendition of *Blackbird* you have ever heard. It is a pity that I was strumming

> the guitar during the filming because it does drown the butcherbird out a bit. However, at the time we thought that the guitar had attracted him. On reflection, though, I am convinced he was just telling us he was wanting a handout. The mother magpie brings her babies here to introduce them every year and of course they get stuck into the seed, etc., which I put out. The baby magpies have a very distinctive cry when they want their mother to feed them. One of the butcherbirds has copied this cry and now will sit on the chair on the verandah and cry like a magpie baby when hungry.[9]

This interspecies 'trying on for size' (Walser 1992, p. 198) resonates with philosopher Vinciane Despret's 'unexpected affinities', when 'animals are invited to other modes of being, other relationships, other ways to inhabit the human world' in order to provoke human beings to address them differently (2010, n.p.).

Two: the confrontation, or coproducing fieldwork

Synopsis

A zoömusicologist looks for productive (and coproductive) ways of being with pied butcherbirds as they sing.

Notes

The rising action of a play's second act typically requires the protagonist to learn new competencies to arrive at a fresh sense of self – such is fieldwork (as well as the explorative activity of PR). Character development cannot be achieved without mentors and co-protagonists. What can we only know together? In conducting my fieldwork, I acted intuitively on participatory action researcher Peter Reason's premise: 'Living as part of the whole starts from the essential insight that we are already participants: we are part of the cosmos, always in relation with each other and the more than human world', even though the wounds of dualism have damaged this (2005, p. 39).

My fieldwork takes in three regions – desert country, saltwater country, and savannah country. I also correspond with and travel to meet a number of people with pied butcherbirds in their lives, and my studies have benefitted from private recordings and videos and the generosity and insights of the recordists who made them. A number of my current field sites have developed from these contacts. The best material inevitably arises out of developing personal relationships (another parallel with PR). I try to never ask for too much, and I never 'take the recording and run'.

My ethical concerns and analytical questions are founded on charity – on asking the best questions possible (see Lestel 2007b, p. 34; Despret 2012). Nonetheless, unlike participatory development, there will be no 'Ward Committees, Stakeholder Forums and User Platforms' convened with songbirds (Nash 2013, p. 110). Neither can my relational expertise or any 'insight into the user's . . . mental model of the tasks' be enhanced via a verbal interview with a pied butcherbird

(Weinberg and Stephen 2002, p. 238). Understanding full well that the location of power in the research process is key to whether it can be considered participatory or conventional (Cornwall and Jewkes 1995, p. 1667), I looked to other animal researchers for novel ways to address and chronicle the problem of representation.

A multi-species ethnography of horse-human relationships reports: 'No horses were interviewed in our study; it is their humans that speak on their behalf' (Maurstad et al. 2013, p. 324). These riders describe their state of co-being with horses as 'intercorporeal moments of mutuality' – entanglements of agentive individuals that serve to co-shape and co-domesticate one another (Maurstad et al. 2013, p. 324). Neuroscientist Gregory Berns also bumped up against issues of animal sentience and permission. He trained dogs to enter an MRI scanner awake and unrestrained: '[W]e treated dogs as persons. We had a consent form, which was modelled after a child's consent form but signed by the dog's owner. We emphasized that participation was voluntary', he continues, 'and that the dog had the right to quit the study. We used only positive training methods' (2013, n.p.).

Human-human encounters also confront the challenge of representation. A case in point: a participatory design study of post-stroke patients enlisted clinical caregivers rather than patients, who suffer 'too many deficits and recovery challenges to interact' with 'a complex cyber-physical machine of considerable size and weight' (Threatt *et al.*, 2014, p. 676). More-than-human research can also take encouragement in computer scientist Jon Whittle's observation that a ' "lean" participatory process can still achieve good outcomes', thus challenging any suggestion that more participation is better (2014, p. 121). Given such diverse styles of representation and coproduction in human-human exchanges, there seems no practical reason why animals would be precluded from PR. Admittedly, I do not set out to *equalize* power relations, but neither do I have total control of the agenda.

Birds have ways of communicating to me their desired degree of participation. Individuals have the opportunity to negotiate and even opt out of my fieldwork, and in several instances this has happened. For example, I was attacked two years running by one bird-musician during the nesting season, although in each instance I managed to escape before taking a direct hit. (Several species of Australian birds are known to vigorously drive off humans if they approach nestlings.) I concluded that s/he was quitting the project, and I no longer record there. Also during nesting season, a group of about eight pied butcherbirds evicted me from their territory via harsh calls and beak claps. I had spent many hours recording there through the years and was undecided whether I had interpreted their message correctly. So, I returned and stood behind a tree, only to be immediately and severely taken to task by a bird who landed onto the trunk just above my head and uttered a loud, harsh call. (While our understanding of the meaning of birds' *songs* is incomplete at best, their *calls* possess semantics; these messages can be social, environmental, identificational, and/or locational, and some, like this, are known to be transspecific.) I swiftly departed.

In reflecting on where zoömusicology sits on a continuum from 'empty rituals of public consultation to complete citizen [or avian] control' (Eubanks 2009, p. 115), I asked myself: What if all pied butcherbirds decided not to participate in my fieldwork – would I don protective gear and soldier on? I would not! I would seek out another project, since I understand my task to be, much like an anthropologist's, 'not to eliminate all disturbance but to disturb well' (Chrulew 2014, p. 32).

Beyond wildlife recordists and songbirds, my mentors and co-protagonists broaden to include fellow campers who have shared anecdotes with me. Accounts from everyday people in everyday language are not so distant from the natural sciences as we are led to believe. Darwin wrote in an accessible style, argued with a personal voice, and imbued his science texts with vivid descriptions, humor, doubt, ambivalence, and human–animal metaphors devoid of scare quotes. His anthropomorphic language springs from the continuity he understands between animals and humans, as well as from his choice of the individual as his unit of currency. Participatory and zoömusicological approaches are consistent with Darwin's use of language, and both reject disinterested research. True, science tends to favor what is most straightforwardly measured and emphasizes the generation of evidence in the form of quantitative data (e.g. Wemelsfelder 1997, Wemelsfelder et al. 2009). However, Darwin made inductive use of anecdotal data to evidence otherwise intractable phenomena (Crist 1996, p. 35).

In addition, citizen scientists excel in ornithology; some possess an expertise that equals or surpasses official scientific knowledge. My research also benefits from the burgeoning field of ethno-ornithology (a branch of ethno-biology), which reports on our relationship with birds and even plays a role in conservation efforts. 'Since far more ordinary people than scientists observe animals', philosopher Bernard E. Rollin stresses, 'it would be a pity to rule out anecdote, critically assessed, as a potentially valuable source of information and interpretation of animal behavior' (1997, p. 133). The more I learn about pied butcherbirds, the better I am able to assess and synthesize accounts from laypersons.

Entr'acte

Ornithologist S.A. White recounts this story during a 1913 scientific expedition in Central Australia:

> The first night we reached a small soakage where some mission natives were stationed to keep the water open for the stock. . . . Our attention was drawn to a black-fronted butcher bird (Cractieus nigrogularius) [sic], commonly called a 'jackeroo', which seemed very tame, and after a while came into the camp for food. One of our boys, Jack, was instructed to ascertain from the natives at the soak why this bird was so tame, and had they made it a pet, and this is what Jack said – 'Him Christian bird, that fellow; him go along a church and pray'. We remembered that the Rev. Mr. Strehlow told us that there was such a bird at the mission, which would persist in going to church on Sundays, and made such a great noise that the preacher's voice could not be heard, so the bird was banished to the scrub around the soak, and this is how we met with what Jack designated 'the Christian bird'.
>
> (White 1914/1998, pp. 108–109)

'It's a fantastic story!' John Strehlow (the Reverend's grandson) told me. 'This has to be the only bird ever exiled for going to church'.

Three: the resolution, or coproducing outcomes

Synopsis

A zoömusicologist and pied butcherbirds coproduce musical outcomes to mutual benefit.

Notes

My outcomes include transcriptions and analyses, to be sure, but my research also hinges on the coproduction of sonic outcomes – or is this a preposterously ambitious conceit? Just as folk tunes and food plants have been refined by countless trials, so too have pied butcherbirds subjected their songs to untold candidate solutions. Thus, as a violinist/composer I do more than *incorporate* avian vocalizations into my practice: I trust the musicality of pied butcherbird song, and many of my (re)compositions are almost direct transcriptions. My ability to transcribe pied butcherbird vocalizations improved by playing them on the violin – with me entering into the physicality of the experience. This became for me part of the analytical process and not merely what preceded or followed it. I study pied butcherbird vocalists, but I also study *under* them.[10]

Identifying outcomes and quantifying impact are critical focal points for PR. In her study into the role of arts and cultural development in contemporary society, Elaine Lally argues that quantitative measures overlook the significant qualitative evidence of impact (2009). Others have argued for PR to rehabilitate the anecdotal over the quantitative report, and to rely on the strength of relationship and narrative over statistical modes of representation and considerations of commercial value.[11] However, qualitative indicators have a habit of morphing into quantitative ones. My computer desktop is currently crowded with forms on how to gather a portfolio of evidence for my creative works: awards, concert attendance, external funding, journal ranking, and citation indices of output spinoffs – my university asks me to justify the contribution and significance of my creative outputs, preferably in numbers and metrics. Participatory geographers Rachel Pain, Mike Kesby, and Kye Askins claim that impact is non-linear and 'can occur throughout research processes', and 'not just from research outputs' (2011, p. 186). They argue that the 'scale of engagement does not equate to quality of impact', which I found helpful and encouraging (2011, p. 186). I never intended to work at a research factory, since 'music is made up of much more than what can be measured' (Avanzini et al. 2003, p. xii).

My birdsong lecture/concerts bear resemblance to an exhibition of artworks organized by an anthropologist on behalf of the people she studies. These *concerts des refusés* allow songbirds to reanimate the art and craft of music-making and its analysis. I contextually situate my solo violin presentations by performing with field recordings. The result is more than a musical object; it's a lived experience. In addition, aesthetic encounters in the concert hall can potentially reset the audience's perception of songbirds, as well as to reconfigure our feelings of responsibility towards our extraordinary planet and those who make their lives here in this time of climate change and biodiversity loss. This is the gist of what I write on grant applications – but must these concerts be framed as instrumental? Perhaps it suffices for the birdsong concerts just to be.

And pied butcherbirds – how, if at all, do they benefit from these concerts? I can only speculate on whether there exists 'aspiration for change' in animals (Bergold and Thomas 2012, §9); at minimum, I would expect a penchant for *no* change – a preference for the status quo over a deteriorating environment. Since climate change is human-made, an encounter with animals that might address this can withstand a somewhat lopsided coproduction, since 'both sides benefit' (Bergold and Thomas 2012, §9). Indeed, researchers have a mandate to proceed.

Marginalized voices: critical reviews

For some, songbirds are a problem demanding empirical information to be gathered, assembled, and analyzed by scientists, while others conceive of them as more than a utilitarian resource – a means to an end. Haraway understands how our conflicting narratives illustrate the significant but ambiguous role that animals and their domination have played in our hierarchical dualisms, writing: 'On the one hand, they were plastic raw material of knowledge, subject to exact laboratory discipline. They could be used to construct and test model systems for both human physiology and politics. On the other hand', she explains, 'animals have continued to have a special status as natural objects that can show people their origin, and therefore their pre-rational, pre-management, pre-cultural essence' (1991, p. 11). Neither object nor toy, an animal is, at least for some, first and foremost a *presence* (Lestel 2007a, p. 100).

Intellectual, political, and economic power and prestige are at the heart of the exclusive authority of Western science to make pronouncements about the more-than-human world, much like that of Western art music concerning the entire sonic experience. The 19th-century strophes of *creativity, originality*, and *genius* cloud the fact that the social glue of music is its most significant trait. Some musicologists continue to insist on the existence of autonomous music, while an everyday, every (wo)man social understanding of music aligns with the participatory emphasis on process and allows for multiple musical meanings.

Zoömusicologists have zoöpolitical work to accomplish. Striking down the 'elite model of power/knowledge relationships' (Lestel 2007a, p. 100) – in this case, those of human exceptionalism in music – can move us more generally away from our human-centered mindset and the flattening of the more-than-human into machines. Zoömusicologists encounter individuals where others see species, exceptions where others see rules, and nuance where others see black and white. They uncover impressive musical activities that scientists gloss over on their mission to explicate function. They demonstrate how previously disenfranchised voices can bring epistemological weight, participatory credibility, and transformative possibility to a project. In a desire to tell a richer story than has yet been told on the capacities of animals (which will in turn illuminate those of our own species), zoömusicologists will relocate, analyze, and coproduce *music* across classic but illusory divides.

Zoömusicology's playhouse thanks a number of unnamed pied butcherbird research partners.

Notes

1 The natural sciences are slowly acknowledging how citizen scientists and volunteers, along with crowdsourcing, remote cameras, and other technologies, might augment animal observations on both time and geographic scales (e.g. Hecht and Cooper 2014). Nonetheless, high-ranking scholarly journals typically favor laboratory studies over observational ones, so theoretical majoritarian texts at the top of the pyramid structure may eclipse meaningful work at the 'bottom'.

2 For more on zoömusicology, see Mâche (1997), Martinelli (2009), Taylor and Lestel (2011), and Taylor (2017). Also see www.zoömusicology.com.

3 Like many music therapists, Guylaine Vaillancourt (2012) believes that Community Music Therapy can contribute to social justice. Music therapist Sue Baines posits the term 'Anti-Oppressive Practice', which underlines her call for 'respectful, efficacious ethically, accountable music therapy' (2013, p. 1).

4 See, for example, Smith-Marchese (1994), Stige *et al.* (2010), Bolger (2013) and McFerran *et al.* (2016).

5 This insight is also indebted to Haraway (1988, p. 581).

6 For more on this and pied butcherbirds in general, see Taylor (2008) and Taylor (2017).

7 I found Pain *et al.* (2011) and Mason *et al.* (2013) particularly helpful in this area.

8 Critics of development strategies, participatory or not, have questioned the sincerity, ease, practicality, and possibility of relinquishing power (see Cooke and Kothari 2001).

9 According to Steve Bell (personal communication, 4 September 2013).

10 However, I do not 'jam' *with* or *at* birds in the field because a singing bird also needs to listen – to conspecifics, to and for predators, and, indeed, to the entire soundscape. I am not interested in placing additional demands on a vocalist but rather in recording what birds in the wild choose to sing without my interference. However, ornithologist and artist Vicki Powys was able to provoke an interspecies 'jam session' with a group of pied butcherbirds (personal communication, 18 September 2008), and others have written to me about similar experiences.

11 For principal-based and regulatory-based versus relationship-based approaches, see Banks *et al.* (2013).

References

Agawu, K., 1995. The invention of 'African rhythm'. *Music Anthropologies and Music Histories*, 48 (3), 380–395.

Aigen, K., 2012. Social interaction in jazz: implications for music therapy. *Nordic Journal of Music Therapy*, 22 (3), 180–209.

Avanzini, G., Faienza, C., Minciacchi, D., Lopez, L., and Majno, M., 2003. General foreword. *In:* G. Avanzini, *et al.*, eds. *The neurosciences and music, vol. 999*. New York: The New York Academy of Sciences, xi–xii.

Baines, S., 2013. Music therapy as an anti-oppressive practice. *The Arts in Psychotherapy*, 40 (1), 1–5.

Banks, S., Armstrong, A., Carter, K., Graham, H., Hayward, P., Henry, A., Holland, T., Holmes, C., Lee, A., McNulty, A., Moore, N., Nayling, N., Stokoe, A., and Strachan, A., 2013. Everyday ethics in community-based participatory research. *Contemporary Social Science*, 8 (3), 263–277.

Beecher, M.D., 1996. Birdsong learning in the laboratory and field. *In:* D.E. Kroodsman and E.H. Miller, eds. *Ecology and evolution of acoustic communication in birds*. Ithaca, NY: Cornell University Press, 61–78.

Bergold, J., and Thomas, S., 2012. Participatory research methods: a methodological approach in motion. *Forum: Qualitative Social Research*, 13 (1), Art. 30.

Berns, G., 2013. Dogs are people, too [online]. *The New York Times*, 5 October. Available from: http://www.nytimes.com/2013/10/06/opinion/sunday/dogs-are-people-too.html [Accessed 17 May 2016].

Bolger, L., 2013. *Understanding and articulating the process and meaning of collaboration in participatory music projects with marginalised young people and their supporting communities*. Thesis (PhD). University of Melbourne.

Chrulew, M., 2014. The philosophical ethology of Dominique Lestel. *Angelaki*, 19 (3), 17–44.

Cooke, B., and Kothari, U., eds., 2001. *Participation: the new tyranny?* London: Zed Books.

Cornwall, A., and Jewkes, R., 1995. What is participatory research? *Social Science and Medicine*, 41 (12), 1667–1676.

Crist, E., 1996. Darwin's anthropomorphism: an argument for animal-human continuity. *Advances in Human Ecology*, 5, 33–83.

Crist, E., 2006. 'Walking on my page': intimacy and insight in Len Howards' cottage of birds. *Social Science Information*, 45 (2), 179–208.

Currie, A., and Killin, A., 2015. Musical pluralism and the science of music. *European Journal for Philosophy of Science*, 6 (1), 1–22.

Dankoff, J., 2011. Toward a development discourse inclusive of music. *Alternatives: Global, Local, Political*, 36 (3), 257–269.

Dennett, D.C., 1987. *The intentional stance*. Cambridge: The MIT Press.

Despret, V., 2010. *Responding and suffering bodies in human-animal worlds* [online]. Available from: http://www.vincianedespret.be/2010/03/responding-and-suffering-bodies-in-human-animal-worlds/ [Accessed 5 January 2015].

Despret, V., 2012. *Que dirait les animaux, si . . . on leur posait les bonnes questions?* Paris: Éditions La Découverte.

Eubanks, V., 2009. Double-bound: putting the power back into participatory research. *Frontiers: A Journal of Women Studies*, 30 (1), 107–137.

Everitt, A., 1997. *Joining in: an investigation into participatory music*. London: Calouste Gulbenkian Foundation.

Gibson-Wood, H., and Wakefield, S., 2013. 'Participation', white privilege and environmental justice: understanding environmentalism among Hispanics in Toronto. *Antipode*, 45 (3), 641–662.

Haraway, D., 1988. Situated knowledges: the science question in feminism and the privilege of partial perspective. *Feminist Studies*, 14 (3), 575–599.

Haraway, D.J., 1991. *Simians, cyborgs, and women: the reinvention of nature*. London: Free Association Books.

Haraway, D.J., 2008. *When species meet*. Minneapolis, MN: University of Minnesota Press.

Hecht, J., and Cooper, C.B., 2014. Tribute to tinbergen: public engagement in ethology. *Ethology*, 120 (3), 207–214.

Hillman, S., 2002. Participatory singing for older people: a perception of benefit. *Health Education*, 102 (4), 163–171.

Iversen, O. S., Halskov, K., and Leong, T. W., 2012. Values-led participatory design. *CoDesign: International Journal of CoCreation in Design and the Arts*, 8 (2–3), 87–103.

Kesby, M., 2005. Retheorizing empowerment-through-participation as a performance in space: beyond tyranny. *Signs: Journal of Women in Culture and Society*, 30 (4), 2037–2065.

Klein, N., 2014. *This changes everything: capitalism vs. the climate*. London: Penguin Books.

Kroodsman, D. E., 1996. Ecology of passerine song development. *In:* D. E. Kroodsman and E. H. Miller, eds. *Ecology and evolution of acoustic communication in birds*. Ithaca, NY: Cornell University Press, 3–19.

Lally, E., 2009. 'The power to heal us with a smile and a song': senior well-being, music-based participatory arts and the value of qualitative evidence. *Journal of Arts and Communities*, 1 (1), 25–44.

Latour, B., 1990. Drawing things together. *In*: M. Lynch and S. Woolgar, eds. *Representation in scientific practice*. Cambridge, MA: The MIT Press, 19–68.

Latour, B., 2005. *Reassembling the social: an introduction to actor-network-theory.* Oxford: Oxford University Press.

Lestel, D., 2007a. *L'animalité*. Paris: L'Herne.

Lestel, D., 2007b. *Les amis de mes amis*. Paris: Éditions du Seuil.

Mâche, F.-B., 1997. Syntagms and paradigms in zoomusicology. *Contemporary Music Review*, 16 (3), 55–78.

Martinelli, D., 2009. *Of birds, whales, and other musicians: an introduction to zoomusicology*. Scranton, PA: University of Scranton Press.

Mason, K., Brown, G., and Pickerill, J., 2013. Epistemologies of participation, or, what do critical human geographers know that's of any use? *Antipode*, 45 (2), 252–255.

Maurstad, A., Davis, D., and Cowles, S., 2013. Co-being and intra-action in horse-human relationships: a multi-species ethnography of be(com)ing human and be(com)ing horse. *Social Anthropology*, 21 (3), 322–335.

McFerran, K. S., Thompson, G., and Bolger, L., 2016. The impact of fostering relationships through music within a special school classroom for students with autism spectrum disorder: an action research study. *Educational Action Research*, 24 (2), 241–259.

Merrifield, A., 1995. Situated knowledge through exploration: reflections on Bunge's 'geographical expeditions'. *Antipode*, 27 (1), 49–70.

Mohan, G., 2001. Beyond participation: strategies for deeper empowerment. *In*: B. Cooke and U. Kothari, eds. *Participation: the new tyranny?* London: Zed Books, 153–167.

Mundy, R., 2009. Birdsong and the image of evolution. *Society and Animals*, 17 (3), 206–223.

Nash, F., 2013. Participation and passive revolution: the reproduction of neoliberal water governance mechanisms in Durban, South Africa. *Antipode*, 45 (1), 101–120.

Pain, R., Kesby, M., and Askins, K., 2011. Geographies of impact: power, participation and potential. *Area*, 43 (2), 183–188.

Parfitt, T., 2004. The ambiguity of participation: a qualified defence of participatory development. *Third World Quarterly*, 25 (3), 537–556.

Plumwood, V., 2009. Nature in the active voice. *Australian Humanities Review*, 46 (May), 113–129.

Reason, P., 2005. Living as part of the whole: the implications of participation. *Journal of Curriculum and Pedagogy*, 2 (2), 35–41.

Rollin, B.E., 1997. Anecdote, anthropomorphism, and animal behavior. *In:* R.W. Mitchell, N.S. Thompson, and H.L. Miles, eds. *Anthropomorphism, anecdotes, and animals*, Albany, NY: SUNY Press, 125–133.

Rolvsjord, R., 2006. Whose power of music? A discussion on music and power-relations in music therapy. *British Journal of Music Therapy*, 20 (1), 5–12.

Smith, R.C., and Kjærsgaard, M.G., 2014. Design anthropology in participatory design from ethnography to anthropological critique? *In: Proceedings of the 13th Participatory Design Conference: Short Papers, Industry Cases, Workshop Descriptions, Doctoral Consortium papers, and Keynote abstracts – Volume 2, Windhoek, Namibia, 6–10 October*. New York: ACM Press, 217–218.

Smith-Marchese, K., 1994. The effects of participatory music on the reality orientation and sociability of Alzheimer's residents in a long-term-care setting. *Activities, Adaptation & Aging*, 18 (2), 41–55.

Stige, B., Ansdell, G., Elefant, C., and Pavlicevic, M., 2010. *Where music helps: community music therapy in action and reflection*. Surrey: Ashgate.

Taylor, H., 2008. *Towards a species songbook: illuminating the vocalisations of the Australian pied butcherbird (Cracticus nigrogularis)*. Thesis (PhD). University of Western Sydney.

Taylor, H., 2011. Anecdote and anthropomorphism: writing the Australian pied butcherbird. *Australasian Journal of Ecocriticism and Cultural Ecology*, 1, 1–20.

Taylor, H., 2013. Connecting interdisciplinary dots: songbirds, 'white rats', and human exceptionalism. *Social Science Information*, 52 (2), 287–306.

Taylor, H., 2017. *Is birdsong music? Outback encounters with an Australian songbird.* Bloomington, IN: Indiana University Press.

Taylor, H., and Lestel, D., 2011. The Australian pied butcherbird and the natureculture continuum. *Journal of Interdisciplinary Music Studies*, 5 (1), 57–83.

Threatt, A.L., Merino, J., Green, K.E., Walker, I., Brooks, J.O., and Healy, S., 2014. An assistive robotic table for older and post-stroke adults: results from participatory design and evaluation activities with clinical staff. *In*: *Proceedings of the SIGCHI Conference on Human Factors in Computing Systems, Toronto, Ontario, Canada, 26 April – May 1*. New York: ACM Press, 673–682.

Townsend, A., 2013. Principled challenges for a participatory discipline. *Educational Action Research*, 21 (3), 326–342.

Vaillancourt, G., 2012. Music therapy: a community approach to social justice. *The Arts in Psychotherapy*, 39 (3), 173–178.

Walser, R., 1992. The polka mass: music of postmodern ethnicity. *American Music*, 10 (2), 198.

Weinberg, J.B., and Stephen, M.L., 2002. Participatory design in a human-computer interaction course: teaching ethnography methods to computer scientists. *SIGCSE Bulletin*, 34 (1), 237–241.

Wemelsfelder, F., 1997. The scientific validity of subjective concepts in models of animal welfare. *Applied Animal Behaviour Science*, 53 (1), 75–88.

Wemelsfelder, F., Nevison, I., and Lawrence, A.B., 2009. The effect of perceived environmental background on qualitative assessments of pig behavior. *Animal Behaviour*, 78 (2), 477–484.

White, Capt. S.A., 1914/1998. *Into the dead heart: an ornithological trip through central Australia*. Adelaide: Friends of the State Library of South Australia.

Whittle, J., 2014. How much participation is enough? A comparison of six participatory design projects in terms of outcomes. *In: Proceedings of the 13th Participatory Design Conference: Short Papers, Industry Cases, Workshop Descriptions, Doctoral Consortium papers, and Keynote abstracts – Volume 2, Windhoek, Namibia, 6–10 October*. New York: ACM Press, 121–130.

3 'Animal–computer interaction: a manifesto' (2011) and sections from 'towards an animal-centered ethics for animal–computer interaction' (2016)

Clara Mancini

Animal–computer interaction: a manifesto[1]

Animals[2] have been involved in machine interactions for many decades. Skinner's famous operant conditioning chamber, used in behavioral experiments since the early 1930s,[3] provided output devices, such as lights or sounds, and input devices, such as levers or buttons, and would dispense food or water if, for example, a rat or a pigeon completed a given sequence of tasks correctly. These systems have gradually evolved into sophisticated computerized environments affording complex interactivity. Other interaction systems, such as computer games currently employed in more advanced primate cognition studies, provide, for example, on-screen animations that can be controlled via joystick (Gill 2011).

Within agricultural engineering, interactive computing devices have also been developed, for example, to optimize milk production in the farming industry, with the introduction of the first automatic milking systems in dairy farms emerging in the early 1990s (Rossing et al. 1997). These systems have rapidly developed into cutting-edge applications of pervasive and ubiquitous computing technology, enabling cows to independently engage in voluntary milking and express intelligent and social behavior never previously observed in constraining farming environments (Brennan 2005).

Examples of a different kind of interaction are provided by tracking and telemetric sensor devices, which have been used in conservation studies since the early 1970s and which have now become commonplace. For example, radio collars allowed researchers to uncover the elusive behavior and territorial needs of snow leopards for the first time (Jackson and Ahlborn 1989), and satellite collars enabled conservation efforts to start mapping the movements of elephants (Lindeque and Lindeque 1991). Tracking devices have also been introduced to the pet market, while various telemetric technologies are used in laboratory settings to monitor, for example, dogs' physiological parameters during pre-clinical trials (Emka Technologies n.d.).[4]

In short, animal–computer interactions have a long history and can be found in many areas in which human activity involves other species.

The elephant in the room

In spite of its history, the study of the interactions between animals and computing technology has never entered mainstream computer science, and the animal perspective has seldom informed the design of animal computing applications, whose development has so far been driven by academic disciplines other than computer science or by other industrial sectors. The design of these technologies remains fundamentally human centred, and the study of how they are adopted by or affect their users remains fundamentally outside the remit of user-computer interaction research.

The negative effects of this lack of animal perspective become obvious when, for example, the behaviour and welfare of seals fitted with bio-logging tags and satellite transmitters are significantly affected and data gathered during costly conservation studies risks invalidation (Hazekamp et al. 2009), or when cows who do not engage with milking systems are culled and farmers suffer capital losses (Brennan 2005). But risk mitigation aside, what about the things we could gain from a shift in perspective? What would it allow us to learn about and achieve with interactive technology? How would it influence our reflection on usability, adaptation, appropriation, methodology, and ethics, to name but a few aspects? Studies in interspecies computer interaction have started making appearances at HCI venues (Lee et al. 2006, McGrath 2009, Noz and An 2011, Weilenmann and Juhlin 2011), but the remarkably marginal position this research still occupies in the HCI community and its research agenda is an indicator that its significance has not yet been recognized. For some reason, animal–computer interaction (ACI) is, quite literally, the elephant in the room of user-computer interaction research. The time has come to acknowledge the elephant, to start talking about ACI as a discipline in its own right, and to start working toward its systematic development.

The right moment

Advances in our understanding of animal and comparative cognition, as well as those in computing technology, make the development of ACI as a discipline both possible and timely, while pressing environmental, economic, and cultural changes make it desirable.

From long-held training experiences, we know that several species can use interactive devices of one kind or another, sometimes appropriating them in interesting and unexpected ways. More generally, though, we now know that many species have sensory faculties superior to ours (Willis et al. 2004), possess sophisticated cognitive abilities, engage in advanced problem solving, use purpose-built tools for complex tasks (Emery and Clayton 2004), communicate through articulated languages, experience a range of emotions, form complex social relationships, make moral judgements (Bekoff 2004), and hand down cultures through generations (Rendell and Whitehead 2001). This has progressively made us more aware of the similarities between humans and other species, more appreciative of other species, and more attentive toward the significance of our relationships with them and the fragile environment we all share (Hurn 2011).

At the same time, the interaction modes afforded by computing technology have expanded well beyond those provided by keyboard and mouse. Tangible, embodied, and proxemics interactions, for example, have brought physicality back into computing by engaging the whole body through contact and movement. Sensor technology has become more agile, robust, and sensitive, better able to read the changes coming from within and around us. In general, developments in pervasive, ubiquitous, and ambient computing are enabling technology to adapt to our spontaneous behaviours and to the contexts that these continuously produce and modify. Not only do these advances make computing technology more accessible to humans but they also make it far more accessible to other species.

Aims and approach

ACI aims to understand the interaction between animals and computing technology within the contexts in which animals habitually live, are active, and socialize with members of the same or other species, including humans. Contexts, activities, and relationships will differ considerably between species, and between wild, domestic, working, farm, or laboratory animals. In each particular case, the interplay between animal, technology, and contextual elements is of interest to the ACI researcher.

ACI aims to influence the development of interactive technology to:

- improve animals' life expectancy and quality by facilitating the fulfilment of their physiological and psychological needs (technology that encourages healthy feeding habits in domestic animals or allows them to modify their housing conditions at leisure might be consistent with this aim);
- support animals in the legal functions in which they are involved by minimizing any negative effects and maximizing any positive effects of those functions on the animals' life expectancy and quality (technology that gives farm animals control over the processes in which they are involved, produces no side effects on the animals involved in conservation studies, or helps working animals communicate with their assisted humans might be consistent with this aim); and
- foster the relationship between humans and animals by enabling communication and promoting understanding between them (technology that allows companion animals to play entertaining games with their guardians or enables guardians to understand and respond to the emotions of their companion animals might be consistent with this aim).

ACI aims to develop a user centred approach, informed by the best available knowledge of animals' needs and preferences, to the design of technology meant for animal use. It also appropriately regards humans and other species alike as legitimate stakeholders throughout all the phases of the development process.

Ethical principles

ACI takes a non-speciesist approach to research (Dunayer 2004), and researchers have a responsibility to:

- acknowledge and respect the characteristics of all species participating in the research without discriminating against any of them;
- treat both human and nonhuman participants as individuals equally deserving of consideration, respect, and care according to their needs;
- choose to work with a species only if the intent is to advance knowledge or develop technology that is beneficial or otherwise relevant to that particular species;
- protect both human and nonhuman participants from physiological or psychological harm at all times by employing research methods that are non-invasive, non-oppressive, and non-depriving;
- afford both human and nonhuman participants the possibility to withdraw from the interaction at any time, either temporarily or permanently; and
- obtain informed consent to the involvement of both human and animal participants, either from the participants themselves (for example, for adult humans) or from those who are legally responsible for them (for animals).

Widespread benefits

The development of ACI as a discipline could have many benefits for both animals and humans. For example, it could have important effects on our interspecies relationships by informing the design of technology that enables the animals we live and sometimes work with to effectively communicate with us, increase their participation in our interactions, and constructively influence our environments. These developments could give us a better understanding of those we share our lives with and help us build safer, richer, longer, and more productive relationships with them.

ACI could also lead to further insights into animal cognition – for example, by informing the design of interactive technology for behavioural studies that affords optimal usability and creative appropriation for the animals. Or it could support conservation efforts – for example, by informing the design of monitoring devices that minimize the impact on the animals while maximizing the quality and reliability of the data gathered through them. Moreover, ACI could improve the economic and ethical sustainability of food production – for example, by informing the design of technology that affords farm animals more freedom and autonomy, enabling them to live less unnatural lives, reducing their stress levels and susceptibility to illness without recourse to drugs, increasing their productivity, and improving the quality of their produce.

Finally, ACI could expand the horizon of user-computer interaction research by pushing our imagination beyond the boundaries of human-computer interaction. For example, it could help us discover new ways of eliciting requirements from those who cannot communicate with us through natural language or abstract concepts. It could help us explore new modes of interaction for those who do not have hands, cannot decipher the patterns emitted by a screen, or have limited attention spans. Or it could help us find new ways of understanding and evaluating the impact of technology on individuals and social groups – perhaps shedding new light on issues such as identity, privacy, or trust, and contributing to our understanding of what it means to be human and who we are in relation to other species.

A research agenda

Of course, whether ACI can yield the benefits outlined here depends on our ability to tackle some challenging questions. For example, how do we elicit requirements from a nonhuman participant? How do we involve them in the design process? How do we evaluate the technology we develop for them? How do we investigate the interplay between nonhuman participants, technology, and contextual factors? In other words, how are we going to develop a user-centred design process for animals? Here is a possible roadmap:

- First, we could look at what has been done in other areas, what knowledge about animal behavior and psychology is available, and what data has already been collected about ACI. We could look at how all that maps onto what we know about user-computer interactions and how it might contribute to ACI as a discipline and design practice.
- Second, we could form collaborations with researchers from disciplines such as ethology, behavioural medicine, animal psychology, and veterinary, agricultural, and environmental engineering to help us with this mapping effort. Similarly, the expertise and experience of professionals and practitioners who work with animals in environments where ACI take place would be important.
- Third, we could study in-the wild cases of whatever technology is already in use or might be developed in order to understand those domains and contexts, their users, and their stakeholders, so that we can begin to develop or adapt relevant ACI concepts and models.
- Fourth, we could look at human-centred interaction design protocols and methods to assess which ones may or may not be relevant to an animal-cent[e]red design process, which might be adapted, which might be borrowed from other disciplines, and which might need to be developed from scratch.
- Fifth, we could start adapting, developing, and integrating ACI design protocols and methods – for example, for requirements elicitation, participatory design, and contextual evaluation, in a loop between empirical work and theoretical reflection.
- Sixth, we could start developing theoretical models of ACI, which would then drive further research. These would take into account pre-ACI research on animals and would be informed by ACI empirical research with animals.

An invitation

Because of the questions it raises and the challenges it poses, ACI is arguably the next frontier in the study and development of interactive technology. Those who are keen on joining in the exploration of this new territory are warmly invited to sign the ACI Manifesto and join our animal–computer interaction group at: http://www.open.ac.uk/blogs/ACI/

****

Selection from 'towards an animal-centered ethics for animal–computer interaction'[5]

Ethical implications of user-centered design for and with animals

In contrast to the above, I argue that, in order to be consistent with a user-centered perspective, ACI's ethical approach to research needs to meet different criteria. First, it is not the animal characteristics that provide grounds for their treatment but rather their role as users and research participants. Thus, giving all animals involved in ACI research equal protection and care (according to their individual needs) is the most appropriate way to ensure that their requirements as users can emerge during the process and can therefore be designed for with their active participation. On the other hand, precisely because user characteristics are so central to the design process, animals cannot be viewed as the substitutable components of an experimental set-up. Therefore, it is only appropriate to involve an animal in research if this is directly relevant to them. Furthermore, if one recognizes that maintaining good welfare at all times is an important individual requirement, in order to be consistent with user-centered design ACI research needs to be compatible with the welfare of both end users and research participants. Protecting the welfare of animals used in research is the aim of related institutional directives, protocols and guidelines. But what are the specific implications of ACI's animal-centered perspective in this regard?

A welfare-centered ethics

What constitutes good welfare for animals is the object of on-going research (Fraser *et al.* 1997, Fraser 2008) some notions of animal welfare assuming more than others that animals are capable of conscious and sentient experience. Because it bypasses the thorny issue of consciousness and sentience, and is therefore relevant to all animals, the notion of welfare proposed by Stamp Dawkins (2012) is particularly useful here. For Stamp Dawkins, animal welfare presupposes the fulfilment of two fundamental conditions: That an animal is healthy and that they have what they want. The author's rationale is that animals have evolved adaptations for coping with environmental conditions (e.g. a thick coat), for exploiting available resources (e.g. specific hunting techniques or a specialized digestive system) and for recovering from injury (e.g. mounting an immune response), in order to maintain good health thus maximizing their chances of survival and reproduction. However, as Stamp Dawkins points out, animals have also evolved adaptations for preventing the occurrence of conditions that could compromise their survival in the first place, adaptations which result in the animals wanting certain things: for example, wanting to search for prey that might be hiding in the ground, or wanting to burrow to hide from potential predators. For Stamp Dawkins, the animal being healthy and having what they want are interdependent conditions or requirements (e.g. a captive animal whose exploratory behavior is constantly frustrated may develop harmful stereotypies) which need to be satisfied at the same time (e.g. giving an animal free access to food needs to be compatible with maintaining their optimal weight). If only one of the two conditions is satisfied, welfare is compromised.

It follows that ACI research should never threaten the health of the animals involved and never deny them what they want, unless denying or limiting what they want is necessary to preserve their health. More specifically, the welfare requirement that an animal is healthy means that ACI research should never entail practices or procedures that interfere with the evolutionary adaptations that support the animal's health (e.g. through genetic manipulations), or threaten the animal's health by compromising their physiological or psychological integrity (e.g. through invasive, aversive, or otherwise injurious manipulations). On the other hand, the welfare requirement that animals have what they want means that ACI research should never entail practices or procedures which prevent animals from expressing spontaneous behavior (e.g. through restriction or constriction), or confine animals within settings that are not those for which they have evolved (e.g. through caging). The only cases in which such practices or procedures would ever be appropriate in connection with ACI research is in the unlikely event that they needed to be carried out for the direct benefit of the individual animal in question (e.g. through therapeutic surgery or confinement).

Stamp Dawkins (2012) notes how generally the death of an animal is not in itself considered a welfare issue on the grounds that a dead animal cannot suffer, from which would follow that the killing of an animal upon completion of a research procedure, an accepted practice by current legislation, does not impact on their welfare unless it causes the animal to suffer in the process. However, such a position seems to be at odds with the very evolutionary definition of animal welfare. If an animal has evolved certain adaptations precisely because these allow him to stay alive, and if violations to the animal's adaptations impact on his welfare, then interventions that lead to the animal's death arguably pose a welfare issue on the grounds that they are incompatible with the very function that has allowed those adaptations to evolve. Bekoff (2010) argues how the struggle of an animal who is under attack indicates that his life matters to him; this point is arguably valid whether the animal is or is not aware of the attack, or even whether he is or is not aware of being alive. As Stamp Dawkins (2012) points out, struggling (e.g. to break free from confinement) is an evolutionary adaptation ultimately aimed at keeping the animal alive and well. Whether the threat to an animal's life is delivered overtly in a form that the animal is able to recognize as a threat (e.g. strangulation) and thus respond to (e.g. struggling), or covertly in a form that the animal is unable to recognize (e.g. lethal injection during sedation) and thus respond to (e.g. hiding), the fact remains that such a threat opposes the very function of the animal's life-preserving adaptations. In this respect, it could be argued that killing can never be compatible with animal welfare, except when the mechanisms whose function is to keep the animal alive and well are so irretrievably compromised (e.g. because of illness) that there is no hope for his health and contentment to be restored to balance. Thus, on welfare grounds, the killing of participating animals at the end of research procedures is incompatible with ACI's animal-centered perspective.

Instead, consistent with Stamp Dawkins' definition of welfare (2012) researchers should always endeavor to respect the animal's identity and safeguard her integrity, both physiological and psychological, at all times. This means that researchers should work in contexts that are habitual for and thus familiar to the animal; they should endeavor to be as unobtrusive and undisruptive of the animal's daily life patterns and routines as possible; they should give the animal space for expression and control over the research process; and they should use only forms of interaction which are respectful of and responsive to the animal's needs and wants at all times. In animal-centered research, the interests of individual participants should 'prevail over the interests of science and society, where there is conflict' (Medical Research Council 2004) and any potential risks to individual participants should outweigh any potential benefit to others. Therefore any cost-benefit analysis of the research should be carried out from the perspective of what, at the best of the researchers' knowledge, are the animal's best interests. In user-centered design this is both and ethical imperative, as recognized by ethics frameworks regulating the involvement of humans in HCI research (Association for Computing Machinery 1992) and a methodological necessity, as argued by Ritvo and Allison in their discussion of research methodologies applicable to ACI (Ritvo and Allison 2014). But how can researchers ensure that, in the case of animals, the interests and requirements of users and research participants are appropriately represented and thus prioritized?

The issue of consent

Existing frameworks motivate the need to minimize the impact of research procedures on the welfare of the animals involved, on the grounds that they are capable of suffering whilst being incapable of consenting. This implies the ability to comprehend the immediate and wider implications of one's involvement (Faden and Beauchamp 1986), but of course interspecies cognitive differences and communication barriers make conveying the welfare implications of a research procedure to other animals practically impossible. Nevertheless, consent arguably marks an important difference between participation and subjection, thus in user-centered research the animals' consent needs to somehow be sought.

Of course, one approach to the issue is seeking consent for animals via mediators who are capable of comprehending the implications of the research in relation to the animals' welfare requirements and who have the legal authority to consent on their behalf. To ensure that consent is provided from a user-centered perspective, such agents should also have a vested interest in prioritizing the welfare of the animals concerned. Furthermore, since in user-centered research participants are not merely representatives of a category or substitutable components of an experimental apparatus but individuals, consent should to be sought on an individual basis. In this regard, Mancini et al. (2012) highlighted the complementary role of the animals' daily carers, on the one hand, and animal welfare experts, on the other hand: The former hold critical contextual knowledge about an individual's characteristic patterns and circumstances, while the latter hold essential expertise to assess those characteristic patterns and circumstances in relation to established animal welfare knowledge (Väätäjä and Pesonen 2013). Thus, overall mediated consent should imply the following:

1 the capacity to comprehend the immediate and wider welfare implications of a procedure,
2 a vested interest in prioritizing the welfare of individual animals,
3 familiarity with the individual's characteristic patterns and circumstances,
4 animal welfare expertise relevant to the individual, and
5 the legal authority to consent on behalf of the animal.

Whether all or part of these competences are distributed across different individuals (e.g. the animal's human companion who is also her legal guardian and an independent animal welfare expert) or are found within one individual (e.g. if the human companion and legal guardian is also an animal welfare expert), they should all be represented in the consenting process. Additionally, an independent authority, such as the animal welfare review bodies envisaged by the European Directive, should ensure that the above conditions are met in compliance with ACI's research ethics framework as well as existing legislation.

On the other hand, voluntary engagement is a fundamental aspect of consent (Faden and Beauchamp 1986); however, clearly it would not be realistic to assume that mediators know what the animal they represent wants in specific contingencies. Thus mediation does not eliminate the need to obtain some form of contingent consent from the animals themselves. While animals might not be able to assess the welfare or wider implications of a procedure, they are nevertheless able to respond to specific conditions (Stamp Dawkins 2012), provided they are afforded the freedom to make relevant choices, including the choice not to engage or withdraw altogether. Ritvo and Allison (2014) propose that participant-controlled procedures are best suited to enable animals' preferences to emerge in ACI research; these may entail dichotomous-choice protocols, whereby participants choose whether or not to be exposed to a stimulus, or multi-stimulus protocols, whereby participants can choose between different stimuli as well as the length of stimulus exposure. If a participant is enabled to choose the pace and modality of their engagement with, or withdrawal from, the research process at any time, then their response can provide a measure of their consent to engaging with a specific research set-up. Of course, any contextual variations during a procedure might affect the participant's assessment of the situation and thus their amenability to participate, so whether participants are able to assess the situation is an important consideration.

Luger and Rodden (2013) argue that, as ubiquitous computing systems become more complex and seamless, and support an increasing range of daily activities, the data that drives their functionalities is increasingly abstracted from its original context; this makes it impossible for (human) users to understand the implications of their interactions with such systems and thus provide informed consent to the use of data they divulge during the course of those interactions. In this respect, the authors emphasize the dynamic nature of consent and the importance of enabling effective withdrawal at any time; they also stress the importance of giving users visibility over data flows within systems and the ability to easily interrogate the system to evaluate the cost-benefit trade-offs of engaging or withdrawing. In a more concrete sense, these are similarly useful considerations when designing ACI research procedures. Thus, overall contingent consent should imply the following:

1. procedural set-ups that enable the animal to assess the situation as much as possible (e.g. allowing the animal to freely explore his surroundings or any research equipment as appropriate prior to starting a procedure, and at regular intervals during the procedure),
2. opportunities for the animal to make relevant choices between alternative forms of engagement (e.g. between different forms of input or output in an interface; between reward mechanisms based on food or play), and
3. the possibility for the animal to effectively withdraw or withhold engagement (e.g. plenty of escape routes or rest corners as appropriate).

Importantly, in order to monitor levels of consent over time, researchers should be able to continually and expertly monitor variations in the participant's response to a procedure against their welfare requirements, as highlighted by Väätäjä and Pesonen (2013), and dynamically and promptly make any appropriate adjustments, including suspending a procedure.

Researchers who work with non-competent or non-linguistic humans (Medical Research Council 2004, 2007) are well familiar with notions of mediated and contingent consent, its dynamic and transient nature (e.g. consent as a process rather than as a one-off occurrence [Medical Research Council 2007]), and the critical importance of monitoring and responding to signs of dissent (e.g. a young child becoming upset [Medical Research Council 2004]). They are also aware of the necessary complementarity of the two forms of consent (Medical Research Council 2004, 2007), whereby those who can see the wider implications of a participant's involvement lend their insight in the participant's best interest, while the participant themselves is the only one who can assess the contingent, directly experiential implications. Consistent with the implications of user-centered research, the very ethical perspective underpinning these notions is just as relevant here.

Notes

1 I am indebted to Yvonne Rogers, Bashar Nuseibeh, Marian Petre, Anne De Roeck, Hugh Robinson, Janet van der Linden, Richard Power, Shailey Minocha, Sandra Williams, Daniel Mills, Shaun Lawson, Helen Sharp, and Simon Buckingham Shum for their constructive criticism and support. This work was supported by the EPSRC grant EP/F024037/
2 The term animal(s) is loosely used throughout to refer to nonhuman animals.
3 Operant conditioning chamber; http://en.wikipedia.org/wiki/Operant_conditioning_chamber
4 [Since original publication this resource is no longer available. See http://www.emka.fr/produit/emkapack4g/ [Accessed 15 May 2016] for a more up to date link.]
5 I am grateful to The Open University's Animal Welfare Ethical Review Body for their comments on this protocol. I am indebted to Kevin McConway for his many thoughtful comments on earlier drafts, to Duncan Banks for his helpful advice, and to Derek Matravers for his encouragement. Alistair Willis and Alma Massaro have taken the time to read and offer precious feedback on recent drafts. This work was funded by The Open University.

References

Association for Computing Machinery, 1992. *Code of Ethics and Professional Conduct*. ACM Council - www.acm.org/about/code-of-ethics (last accessed 02.03.16).

Bekoff, M., 2004. Wild justice and fair play: cooperation, forgiveness, and morality in animals. *Biology and Philosophy*, 19 (4), 489–520.

Bekoff, M., 2010. *The animal manifesto: six reasons for expanding our compassion footprint*. Novato, CA: New World Library.

Brennan, Z., 2005. Ooh-aaar, cows do milk themselves, down on the robotic farm [online]. *The Sunday Times*, 23 October. Available from: http://www.timesonline.co.uk/tol/news/uk/article581764.ece.

Dunayer, J., 2004. *Speciesism*. Derwood, MD: Ryce Publishing.

Emery, N. J., and Clayton, N. S., 2004. The mentality of crows: convergent evolution of intelligence in corvids and apes. *Science*, 306 (5703), 1903–1907.

Emka Technologies, n.d. *Telemetry in primates*. Available at: http://www.emka.fr/telemetry-in-primates-94.html [no longer accessible]

Faden, R., and Beauchamp, T.A., 1986. *History and theory of informed consent*. Oxford: Oxford University Press.

Fraser, D., 2008. *Understanding animal welfare: the science in its cultural context*. Oxford: Wiley-Blackwell.

Fraser, D., Weary, D.M., Pajor, E.A., and Milligan, B.N., 1997. 'A scientific conception of animal welfare that reflects ethical concerns', *Animal Welfare*, 6, 187–205.

Gill, V., 2011. Monkeys 'display self-doubt' like humans [online]. *BBC Earth News*, 21 February. Available from: http://news.bbc.co.uk/earth/hi/earth_news/newsid_9401000/9401945.stm [Accessed 15 May 2016].

Hazekamp, A.A.H., Mayer, R., and Osinga, N., 2009. Flow simulation along a seal: the impact of an external device. *European Journal of Wildlife Research*, 56 (2), 131–140.

Hurn, S., 2011. *Humans and other animals: cross-cultural perspectives on human-animal interactions*. London: Pluto Press.

Jackson, R., and Ahlborn, G., 1989. Snow leopards (*panther unica*) in Nepal: home range and movements. *National Geographic Research*, 5 (2), 161–175.

Lee, S.P., Cheok, A.D., and James, T.K.S., 2006. A mobile pet wearable computer and mixed reality system for human-poultry interaction through the Internet. *Personal and Ubiquitous Computing*, 10 (5), 301–317.

Lindeque, M., and Lindeque, P.M., 1991. Satellite tracking of elephants in Namibia. *African Journal of Ecology*, 29 (3), 196–206.

Luger, E., and Rodden, T., 2013. An informed view of consent for Ubicomp. *In: Proceedings of the 2013 ACM international joint conference on Pervasive and ubiquitous computing, 8–12 Sept, Zurich, Switzerland.* New York: ACM Press, 529–538.

Mancini, C., van der Linden, J., Bryan, J., Stuart, A., 2012. *Exploring Interspecies Sensemaking: Dog Tracking Semiotics and Multispecies Ethnography*. Proc. ACM Ubicomp'12, ACM Press, New York, pp.143–152.

McGrath, R.E., 2009. Species-appropriate computer mediated interaction. *In*: *CHI '09 Extended Abstracts on Human Factors in Computing Systems, 4–9 April, Boston, MA*. New York: ACM, 2529–2534.

Medical Research Council, 2004. *MRC ethics guide: medical research involving children*. London: Medical Research Council. Available from: www.mrc.ac.uk/documents/pdf/medical-research-involving-children [Accessed 15 May 2016].

Medical Research Council, 2007. *MRC ethics guide 2007: medical research involving adults who cannot consent*. London: Medical Research Council. Available from: http://www.mrc.ac.uk/documents/pdf/medical-research-involving-adults-who-cannot-consent [Accessed 15 May 2016].

Noz, F., and An, J., 2011. Cat cat revolution: An interspecies gaming experience. *In: Proceedings of the SIGCHI Conference on Human Factors in Computing Systems, 7–12 May, Vancouver, BC*. New York: ACM, 2661–2664.

Rendell, L., and Whitehead, H., 2001. Culture in whales and dolphins. *Behavioral and Brain Sciences*, 24 (2), 309–382.

Ritvo, S., and Allison, R., 2014. Challenges related to nonhuman animal-computer interaction: usability and 'liking'. *In*: *Proceedings of the 2014 Workshops on Advances in Computer Entertainment Conference, 11–14 Nov, Funchal, Portugal*. New York: ACM, Article No. 4.

Rossing, W., Hogewerf, P.H., Ipema, A.H., Ketelaar-De Lauwere, C.C., and Koning, C.J. A.M.d., 1997. Robotic milking in dairy farming. *Netherlands Journal of Agricultural Science*, 45 (1), 15–31.

Stamp Dawkins, M., 2012. *Why animals matter: animal consciousness, animal welfare and human well-being*. Oxford: Oxford University Press.

Väätäjä, H., and Pesonen, E., 2013. Ethical issues and guidelines when conducting HCI studies with animals. *In*: *CHI '13 Extended Abstracts on Human Factors in Computing Systems, 27 April – 2 May, Paris*. New York: ACM, 2159–2168.

Weilenmann, A., and Juhlin, O., 2011. Understanding people and animals: the use of a positioning system in ordinary human canine interaction. *In*: *Proceedings of the SIGCHI Conference on Human Factors in Computing Systems, 7–12 May, Vancouver, BC*. New York: ACM, 2631–2640.

Willis, C.M., Church, S.M., Guest, C.M., Cook, W.A., McCarthy, N., Bransbury, A.J., Church, M.R.T., and Church, J.C.T., 2004. Olfactory detection of human bladder cancer by dogs: proof of principle study. *British Medical Journal*, 329 (7468), 712.

4 Transformations of time on ecological pilgrimage

Peter Reason

As I write, I have in front of me three pieces of rock from the coast of the northwest Highlands of Scotland. The first is a pebble of pink granite, collected on the beach on the southwest corner of Mull. I had anchored my little yacht Coral in the narrow bay at Rubh' Ardalanish and gone ashore to explore, picking it out from the beach as I returned. It was startlingly pink when I lifted it out of the cold water; now dry it is more subdued but still remains strongly coloured, its coarse crystalline structure evident to the naked eye. Granite is an igneous rock, its crystals formed in intense heat and pressure under the Earth surface, cooling as it bubbles through in dome-shaped formations. My piece of granite erupted some 50 million years ago as part of the major earth movements that formed the Highlands. In geological terms, it is broadly contemporaneous with the columnar volcanic basalt, most celebrated on the small island of Staffa, but generally common across this area.

After exploring Mull, I made my way north, past the Small Isles and Skye to the northwest mainland, finding an anchorage surrounded by the massive rounded hills of Loch Torridon. Here I found my second piece of rock: a pebble, deep browny-red, composed of fine, even particles with no crystalline structure to be seen. This is Torridonian Sandstone, a sedimentary rock more ancient by far than the igneous rocks of Mull and Skye. Composed of the eroded fragments of earlier formations and deposited about a billion years ago as a low relief plain, it has weathered through eons of time into the rounded mountains that characterise this landscape.

My third piece of rock comes from yet further north. It is rough and craggy, with a crystalline structure more coarse even than the granite. On one side, bands of dark and light crystals are scattered with gleaming dots of fools' gold; the other side is pinky-red embedded with fragments of mica. This is Lewisian gneiss, collected from the far northwest. This rock originates in the Precambrian era, up to three billion years old, a metamorphic formation forged by the transformation of even older rock by the enormous heat and pressure as the crust of the primal Earth solidified. These rocks were already ancient and eroded when the sandstone was laid down; they are among the oldest on the planet.

I collected these rocks as I travelled through and dwelled within these geological landscapes, contemplating the ancient rocks, picking over them on the foreshore, reading about them in the geology books in my ship's library (including my favourite, Fortey 2010). And through this dwelling and contemplation, I entered into experiences of time quite different from the clock time of my everyday life.

****

This voyage was part of an experiment in ecological pilgrimage. Over two summers I sailed, mainly single-handed, on the western edge of the British archipelago: from Plymouth, past Land's End and the Scillies, across the Celtic Sea, up the west coast of Ireland, across to Scotland and the Western Isles, and on to the coast of the northwest Highlands.

The idea of pilgrimage draws on many sources. It is part of the tradition, possibly as old as the human species, of making a more or less arduous journey away from the comforts and familiarities of home in search of new insights and deeper understandings. The faithful embark on religious pilgrimages to encounter a holy realm; my ecological pilgrimage took me away from the habits of civilization and disrupted the patterns of everyday life in search of a vision of the Earth of which we are a part. I took as my 'text' Thomas Berry's lament that 'we are only talking to ourselves', that we are no longer talking to the river or the seas and indeed no long listening to them (Berry and Clarke 1991, p. 20).

At these times when human activities threaten the continuation of civilisation and complex life on Earth, I carry with me questions I believe are critical: How can we Western humans learn to understand, emotionally and spiritually as well as intellectually, that we are entirely part of and dependent on the natural world? How can we understand this in a way that is self-evident, utterly natural, for us – in the same way that the existence of God was self-evident in medieval society and the world of objects is to modern humans?

The modern eye may see pilgrimage in its traditional sense as full of superstition, self-delusion and even mass hysteria. However, poet and wilderness writer Gary Snyder describes the wilderness pilgrim's 'step-by-step breath-by-breath' progress into the wild, whether the wild of mountains or ocean or meditation, as 'an ancient set of gestures' that bring a sense of joy, a joy that arises through 'intimate contact with the real world' and so also with oneself (1990, p. 94). Douglas Christie (2013) shows this was also true of the Desert Fathers of the early Christian era: Their contemplative disciplines took them between inner and outer landscapes in search of a consciousness of the whole of creation (see also Reason 2013). So if we are able look beyond modern prejudices to this 'ancient set of gestures', we may discover how practices of pilgrimage might inform the development of ecological sensitivity and responsiveness.

Pilgrimages into the wild world are one response to the ecological crisis of our times; intimate knowledge and appreciation of our home patch is another. Both are ways of restoring our enchantment to the world of which we are a part. Re-enchantment is not, of course, a sufficient response, for we also urgently need a whole range of political, financial, technological, and cultural initiatives that would change society as we know it. I think it is nevertheless a necessary response, one that may inform these more practical and political concerns. Yet, opening oneself to the wild world and describing what one finds with love and passion is itself a political and spiritual act.

My pilgrimage was also a process of inquiry. As a longstanding practitioner and theorist of action and participatory research (Reason and Rowan 1981, Reason and Bradbury 2001, 2008), I have argued that such approaches to inquiry are not simply alternative research methods, but reach toward a participatory worldview that challenges Western dualism. At its fullest ambition this leads to 'living as part of the whole' that places humans in the web of life as embodied participants (Reason 2005). My colleagues developed work along similar lines: Judi Marshall (1999, 2004, 2016) in both personal and professional spheres strove to practice what she called 'living life as inquiry'; Bill Torbert (1991, 2004, Reason and Torbert 2001) developed 'action inquiry', to be conducted in everyday life in the interests of human persons, their communities, and the ecosystems of which they are a part.

A primary vehicle for such inquiry is an attention that encompasses experience, representation, sense-making and action. The development of such attention can be cultivated through mindfulness disciplines such as meditation, martial arts, and the practices of living life as inquiry; sailing the Atlantic coasts single-handed in a small boat helps cultivate a similar quality of attention. This is not to say that such attention is always present: that would be claiming far too much. Rather, we stumble along, often distracted, falling short of our aspirations; as my Buddhist teacher would say, we pull ourselves through life – as through meditation – mistake by mistake by mistake.

My approach to inquiry was intense but simple: I sailed off on my own with notebook, audio recorder, cameras, and iPad. While managing my small boat in more or less wild waters, I quite simply immersed myself in the landscape and seascape, sometimes drawing on formal meditation and Tai Chi practices, sometimes just being there. I kept records of what I saw and felt; from time to time, I wrote and posted more carefully composed blogs (onthewesternedge.wordpress.com). On board, I had sailing directions and other travel guides as well as reference books on wildlife and geology that I used to deepen my understanding of what I was seeing.

In action research language, I was engaged in cycles of action and reflection, moving through an extended epistemology of experience, representation, sense-making and action (Heron 1996, Heron and Reason 2008). The pilgrimage itself was one large cycle (which is of course ongoing as I return home and try to integrate my learning into everyday life); it included smaller cycles during the voyage itself. These cycles were initially 'first person inquiry', but as my experience deepened, from time to time I touched on a world not of things but of presences with which I was required to negotiate. As I wrote in *Spindrift*, my first book of ecological pilgrimage (Reason 2014, p. 125), 'The world around me took on a subjective presence', which led me in and out of a 'second person' inquiry.

The experiences on pilgrimage were then deepened by the narrative writing that I engaged in on my return home. It is written primarily as 'nature writing' or 'ecological literature' rather than in academic form.

****

I have my first intimations of the disruption of my sense of time while at anchor between the low, grassy Inch Kenneth and the huge basalt cliffs at the entrance to Loch na Keal on the west coast of Mull. I approach the sheltered anchorage cautiously: It is guarded by underwater reefs and there are no clear indications of the passage between them. Once safely in and settled, I look around me. Coral is anchored between the hard and the friable. Inch Kenneth is formed of sedimentary rock where ancient conglomerates and limestone outcrops have broken down to give good grassland. The cliffs, in contrast, are volcanic in origin: Sequential eruptions laid down layers of Triassic basalt; over time these eroded into terraces stepping down the hillsides. At the base of the cliffs, by the road along the shore, stand three tiny white cottages. In many ways they are quite insignificant, but also a testament to human ability to create living space in the most unlikely places.

I decide to stay put for a day. It is early in my pilgrimage, I need to settle in, take things gently. The following day it is raining and windy, and I see no point in getting wet and uncomfortable and so stay another night. I devote periods to formal meditation, quietening my mind then opening to the land around me. With the rain spitting and fresher winds rocking the boat, as well as my restless mind, concentration is difficult. In time, however, I am able to really attend to these cliffs. I watch them through the day as the sun moves across the sky, casting shadow in the morning and lighting their peaks with orange in the evening. I notice the details of the streams tumbling down, glimmering where the light catches the falling water. I absorb the contrast between the cottages at the foot and the enormity of the 200–300 metres drop. All this attention provides me with a tiny sense of intimacy, of being in place rather than watching scenery. And I am struck by the contrast between my human impatience, my restlessness to get on, and the simple presence of these cliffs.

Without really thinking about it, I am drawing on a practice of deep participation strongly influenced by the work of the philosopher Henryk Skolimowski, one of the first to articulate the possibility of 'participatory mind'. He argues that Western persons are conditioned by what he called a 'Yoga of Objectivity', a 'gentle form of lobotomy' that teaches us that things exist in isolation. To develop a participatory mind we need training exercises: a 'Yoga of Participation'. This Yoga consists of a series of practices that one can draw on in an encounter with another being. He outlines these as a) preparing one's consciousness by calming the mind; b) meditating on the form of being of the other; c) reliving its past, its present, its existential dilemmas; d) asking permission to engage with it; e) praying to be allowed to enter into communion; f) in-dwelling in compassionate, empathic terms, exploring what forms of dialogue were possible; and g) withdrawing with thanks and gratitude (Skolimowski 1994, p. 147–164).

In this practice, I am glimpsing time at the limits of human imagination. While the basalt rocks and my pink pebble are geologically quite young, they were formed long before the ancestors of the *homo* species emerged. And yet those little cottages appear so permanent, so part of the scene. There is a lesson about the nature of time here, but at this early stage of my pilgrimage I am not yet ready for it. I need to get further into the experience and allow it to disrupt my everyday sense of time more thoroughly.

****

Late one midsummer evening some weeks after my stay at Inch Kenneth, I put aside my book and climb the companionway to the cockpit. Coral is anchored in the pool off Tanera Beg in the Summer Isles, a few miles north of Ullapool. The sun is poised just above the peak of the Eilean Fada Mór to the northwest, throwing a rich golden light onto the sandstone rocks that circle the anchorage – the same sandstone deposit as the dark red pebble I picked up at Loch Torridon a few miles to the south. A heron stands in the deep shadow along the shoreline, motionless, poised to strike. A few gulls cry harshly; there is a twittering of land birds from the shore. The flag halliard rattles lightly against the backstay. Otherwise silence.

Opposite the sinking sun, the three-quarter moon is rising into a just-blue sky over the line of mountains on the distant mainland. The low sunlight highlights the ridges and casts the valleys into shadow, giving the mountains a dimensionality and body even though they are in the far distance.

The sky is clear apart from a few wisps of dark cloud over the peaks: A slack weather system with patchy cloud and light, variable winds has persisted for the best part of a week. The sea reaches calm all the way to the mainland shore, tiny ripples moving dark shadows hypnotically across the surface. The tide is falling, revealing the reefs at the entrance to this pool and uncovering the coral beach, ghostly pale in the failing light, where an oystercatcher is busily hunting along the water's edge.

Night is coming, and yet at this time of year and at this latitude it will be scarcely dark, especially with the near-full moon high in the sky. In the time it takes to scribble a few words in my notebook, the sun has disappeared, the distant mountains seem to be in a greater light; the moon has risen higher and is more clearly defined in a darkening sky.

****

My everyday life is dominated by clock time: I wake and retire, have my meals, arrange to meet people, pretty much to a set schedule. This remains true at the start of my voyage, as I religiously consult the almanac for the times of high and low water for the week ahead and note them down in my tidal atlas. But as the pilgrimage unfolds, while there remains a sense of time passing, this is increasingly marked, not by the digital regularity of clocks, but by the natural rhythms of the planet that I shall call 'Earth time'.

Two months into my ecological pilgrimage, anchored in the Summer Isles, I am saturated in Earth time. This is the first major disturbance to my sense of time. I experience time passing not by one digital metric, but by a series of overlapping rhythms. The sun rises and sets; the moon waxes and wanes; high tide moves forward about 50 minutes each day; I get hungry and I eat, tired and I rest. The passing of weather systems brings a longer beat to the rhythm. When slow moving or slack weather systems predominate – as in the 'long hot days of summer' – little changes to mark the passing of time. But when depressions move in quickly from the Atlantic, bringing fresh, changeable winds, they stir up sense of change and even urgency. Clock time never disappears completely, of course; it simply becomes another strand in the weave, only salient when, for example, I need to know the start of the favourable stream through a narrow passage.

****

As clock time fades in significance and a more direct encounter with the wild world distracts attention from everyday preoccupations, social constructions of reality fade away. This allows for a second experience of time that I call the 'eternal present': those moments when clock time appears to stand still and differences between self and other, inner and outer, disappear. At such, often tiny, moments it is as if there is a crack in the cosmic egg through which a different world opens that is nevertheless the same world.

Leaving Scalpay, it is more of a performance than usual to get out of the harbour and on my way. It takes a while to get the dinghy on board and properly stowed. The anchor chain comes up black and sticky, spreading mud all over the foredeck, taking several buckets of seawater and a lot of scrubbing to clean up. As I motor out of the harbour there are unfamiliar rocks and reefs to negotiate. And once the mainsail is hoisted and set, it is clear that there is very little wind, and to make any progress at all I need to rig the inner forestay and hoist the No.1 genoa – a big sail that sweeps the deck and reaches nearly as far aft as the cockpit – rather than just unfurl the working foresail. For nearly an hour I seem to be constantly on the go from cockpit to foredeck and back again.

Once settled, with all sail set, Coral sails elegantly toward Skye, rippling the unusually smooth waters of the Little Minch, making just over three knots. But soon the wind fades. Coral's speed drops below three knots, then two, and after creeping along for half an hour or so, to nothing at all. 'Let it be', I tell myself. 'There is plenty of daylight, we are not unsafe or uncomfortable'. I allow Coral to drift about in the middle of the sea.

The day is pleasantly warm. Loose cloud covers the sky, the sun shining fitfully through the gaps. The wind is even more fitful, ruffling the water, promising some action, and then fading into nothing very much. Ahead lies Skye, a dark silhouette; astern, bright sunshine picks out the outcrops of gneiss on Harris. Fair weather cumulus rises along the whole line of the Outer Hebrides from Barra to Stornoway. The sea is quiet, undulating like a dimpled mirror, throwing shallow reflections this way and that. Both sea and sky are the same exquisite silvery grey, meeting in the far distance north and south at a horizon that is diffuse and uncertain.

Now I have stopped fiddling around trying to get Coral to sail, I am open to the wonder of the moment. Held in a space between two lands, and with the sea merging into the sky, the sky into the sea, I lose myself into this wider, silvery world. My sense of self becomes as diffuse and uncertain as that horizon. I am still present, but with no sharp distinction between in here and out there, I become part of the quiet presence of the world.

****

As Gary Snyder puts it, such 'sacred' moments take one away from one's little self into the wider whole (1990, p. 94). If I have learned anything in three long seasons of sailing on the western edge, of pilgrimage in search of a different kind of relation to the Earth on which we live, it is that these sacred moments arise quite spontaneously and unexpectedly. They certainly cannot be forced, although I notice they often come when I step back from preoccupation with the demands of sailing and pilotage. But on occasion, they arise in the midst of such preoccupations.

Sailing north, making for Ullapool, in strong and gusting winds, I secure the double-reefed mainsail with a preventer and pole out the genoa with the spinnaker boom. Through the morning Coral blows fast, but not uncomfortably, up the coast toward the headland Rubha Reidh, keeping safely well offshore. I make myself coffee, then soup and a chunk of bread for lunch. Coral passes the headland and it is time to gybe round into outer Loch Broom. (Sailors will know that gybing involves bringing the mainsail across from one side of the boat to the other; done properly, it is a safe manoeuvre; uncontrolled, it can be dangerous.) I start the routine of getting Coral ready for the turn: I roll in the genoa; then, secure with lifejacket and harness, I go forward to the exposed foredeck to lower and stow the spinnaker boom.

It is only when back in the cockpit that I look astern at the following sea. The waves seem bigger than I had expected, the troughs between them deeper. For a moment I hesitate. Can I really do this safely on my own? The grey-green surface of the approaching wave looks cold, relentless and implacable, a small hillside of water, then another, then another. I watch the waves for a few short seconds and my fear drops away. Now, in recollection, I would describe this as a moment of direct meeting, when I am simply present in this wild sea with no thoughts and no self-concern. By watching the waves roll toward Coral's stern, I have tuned myself to their rhythm: Without conscious decision the, moment of action arrives. I haul in the mainsail hand over hand and jam it firm in the cleat. I lean against the tiller with my thigh and hold it there. Coral's stern comes round through the wind and the mainsail, constrained by the tight sheet, flips safely through its short arc. I slowly pay out the sheet so Coral settles onto the opposite tack, driving eastward into the loch. With the mainsail safely gybed, I roll out the genoa again and, after peering through the spray into the distance and comparing what I can see with what was shown on the chart, set a course to pass to the south of Priest Island.

****

In many ways there was nothing special about this. I was on my own in rough and windy weather, but a gybe is just a gybe, one of many on a three-month voyage. But in the weeks that passed, the image of those implacable waves rolling up behind Coral's stern kept returning, until I found some illumination in David Hinton's book *Hunger Mountain*, in which he explores the wisdom of the ancient Chinese poets and sages (2012, p. 105). In contrast to the everyday Western experience of the self as separate from the world and acting intentionally as a 'transcendent spirit centre', Hinton considers the terms *wu-we* and *tzu-jan*. *Wu-we* means 'not acting', in the sense of 'acting without the metaphysics of self'. By being absent or self-less while acting, Hinton suggests that 'Whatever I do, I act from that source and with the rhythm of the Cosmos' (the "I" in the sentence must be read ironically, in the sense of 'this particular being'!). *Tzu-jan* is usually translated as 'suchness', and points to the spontaneous unfolding through which the world burgeons into presence. And while I don't want to get into absurd and unsustainable claims of selflessness, it does seem to me that in the moment of the gybe I was sufficiently tuned to the boat and the wind and the sea that the action was accomplished with an elegance that was not just of my own making.

These tiny moments when time stands still, or maybe more accurately becomes irrelevant, are easy to overlook or to see as insignificant. They are not overwhelming transformations of consciousness, but nevertheless are profoundly important in calling forth a different relationship with the world: no longer out there as landscape, but recognised as a subjective presence of which we are part. The challenge, the creative opportunity, is quite simply to be open to these moments when they arise.

****

Further north again, the sea around Coral turns a bright turquoise blue as I sail past Rubha Coigeach into Enard Bay. The low shore ahead is gnarled and lumpy grey gneiss, with many small inlets and scattered islands. From out at sea I can see the succession of sandstone mountains – Stac Pollaidh, Suilven, and Quinag – rising as a sculptured line above the gneiss platform.

I find a place to stop overnight in the small but sheltered anchorage at Loch Roe. As I close the shore, the gnarled crystalline rocks seem featureless. The sailing directions refer to a high bluff that distinguishes the entrance; but there seem to be high bluffs all along the coast, and I find it difficult to discern the way into the small loch against the background of grey rocks merging into each other. On first approach, I sail past the entrance before I realise my mistake. With the sails down I motor back, close along the shoreline. Ah! that must be the bluff of rock, there are the offshore rocks marked on the chart. I steer closer into what looks like a narrow entrance, ready to retreat at a moment's notice. The little bay ahead ends in a beach strewn with seaweed and plastic litter, but a passage to starboard opens between a tidal island and a patch of floating bladderwrack that indicates the presence of an underwater reef. Beyond is a tiny pool, deep, with just room to swing at anchor. A sheer cliff of crystalline gneiss stands high above the cockpit to one side, a line of rocks and tidal islets where seals are resting provide shelter on the other. And in the distance, the ridge of Quinag rises above the flattened landscape, clearly visible across the top of the loch.

The tide rises so the pool is full to the brim, covering the rocks and chasing the seals away. The swell from the sea finds its way between the islets, moving Coral gently around. Enchanted, I sit in the cockpit as the sun goes down, watching the light play on the rocks and mountains, absorbing an unfathomable sense of geological time. This is the place I have been looking for, where I catch a glimpse of a world both eternal and made anew in every moment.

****

Just what is it about these different rocks and mountains that I find so satisfying to see, to be in the presence of? Apart from the fascination of their different origins; apart from their beauty and grandeur, the way they shapeshift with the changing light; and apart from the contrasting landscapes they give rise to?

It seems to me that beyond all that there is something simply inconceivable about their age, their origins, their history, and that this is an opening of a third dimension of 'deep time'. Confrontation with the age of the Earth – and beyond that the Cosmos – allows a glimpse of time as the container of all possibilities. Dwelling with these mountains – by which I mean spending time with them, meditating on them, studying their origins – gives me some sense of deep time quite different from what I have called Earth time and the eternal present. We can measure their age, but the age we derive is really beyond our grasp: Truly, whose mind can encompass two and a half billion years? We might also consider that, in terms of time passing, the Sun and Earth and solar system are more or less halfway through their lifespan. If we struggle to consider two and a half billion years of rock formation, or four and a half billion years of the Earth's existence so far, how inconceivable is it to think of the evolutionary possibilities that are latent in another four and a half billion years before the whole is swallowed in the red giant that the Sun will eventually become?

Contemplating these issues, I come to think of deep time as close to the Taoist notion of Absence, the pregnant emptiness from which all things appear and to which they return in a process of perpetual transformation. We cannot know this absence directly, but we may get a suggestion of it through the ancient presence of these rock formations and mountains. They seem to point beyond time in any human sense to a timelessness at the heart of existence.

****

A few weeks later Coral and I slip back into the marina and moor safely, just ahead of the stormy weather brought to these islands by the tail end of Hurricane Bertha. I have just enough time to tidy up, pack my clothes and order a taxi that will get me to the railway station in time for my reservation. I have an advance single ticket that gets me home just as I planned many months ago. At the end of my pilgrimage, clock time summons me back to social reality.

Back home, I tell the story of my descent in time to Elizabeth, my wife, passing each of my three pieces of rock to her in turn, explaining where I collected them and their geological origins. She listens intently and responds with delight, telling me that holding the stones helps her connect with my story and my pilgrimage.

I continue my inquiry, reflecting on my experience, making sense and composing a narrative. My reflections on time become more focussed when my friend Bill Torbert sends me his thoughts about time in a draft paper (2014) and I adopt, with amendments, the three dimensions of time he outlines. I write a draft of this chapter, drawing on my memories, reading my notebook and studying my photographs. This is the reflective, sense-making, stage of my inquiry, where I am integrating my story in both presentational and propositional form (Seeley and Reason 2008).

I am invited to talk about ecological pilgrimage and to read from my earlier book *Spindrift* at various literary and ecological events. I take my rocks with me and pass them around as I tell the story of my descent into deep time. People seem to immerse themselves in my story as they hold the rocks: the obloid pink granite pebble that sits smooth and heavy in the palm of the hand; the smaller piece of sandstone that one wants to rub with the thumb while holding in the fingers; the jagged gneiss that demands you turn it over to examine its different facets and run a finger along its sharp edges. It appears there is a deepening sense of engagement with deep time as people hold the rocks. But how would we know?

I place the rocks in the middle of our dining room table, where they stay through the winter, a reminder and stimulus for occasional conversation. As the New Year awakens, they are put away and replaced by vases of spring flowers. As I finish this writing a year and more on, I hunt them out and put them on my desk again: They hold my memories and are a way of bringing my pilgrimage home.

References

Berry, T., and Clarke, T., 1991. *Befriending the earth: a theology of reconciliation between humans and the Earth.* Mystic, CT: Twenty-Third Publication.

Christie, D. E., 2013. *The blue sapphire of the mind: notes for a contemplative ecology.* New York: Oxford University Press.

Fortey, R., 2010. *The hidden landscape: a journey into the geological past.* London: The Bodley Head.

Heron, J., 1996. *Co-operative inquiry: research into the human condition.* London: Sage Publications.

Heron, J., and Reason, P., 2008. Extending epistemology with co-operative inquiry. *In:* P. Reason and H. Bradbury, eds. *Sage handbook of action research: participative inquiry and practice.* London: Sage Publications, 366–380.

Hinton, D., 2012. *Hunger mountain: a field guide to mind and landscape.* Boston, MA and London: Shambhala.

Marshall, J., 1999. Living life as inquiry. *Systematic Practice and Action Research,* 12 (2), 155–171.

Marshall, J., 2004. Living systemic thinking: exploring quality in first person research. *Action Research,* 2 (3), 309–329.

Marshall, J., 2016. *First person action research: living life as inquiry.* London: Sage Publications.

Reason, P., 2005. Living as part of the whole. *Journal of Curriculum and Pedagogy*, 2 (2), 35–41.

Reason, P., 2013. Review, the blue sapphire of the mind: notes for a contemplative ecology by Douglas E Christie. *Resurgence & Ecologist*, 281 (November/December), 60–61.

Reason, P., 2014. *Spindrift: a wilderness pilgrimage at sea.* Bristol: Vala Publishing Cooperative.

Reason, P., and Bradbury, H., eds., 2001. *Handbook of action research: participative inquiry and practice*. London: Sage Publications.

Reason, P., and Bradbury, H., eds., 2008. *Sage handbook of action research: participative inquiry and practice*. London: Sage Publications.

Reason, P., and Rowan, J., eds., 1981. *Human inquiry: a sourcebook of new paradigm research.* Chichester: Wiley.

Reason, P., and Torbert, W.R., 2001. The action turn: toward a transformational social science. *Concepts and Transformations*, 6 (1), 1–37.

Seeley, C., and Reason, P., 2008. Expressions of energy: an epistemology of presentational knowing. *In*: P. Liamputtong and J. Rumbold, eds. *Knowing differently: arts-based and collaborative research methods.* New York: Nova Scotia Publishers, 25–46.

Skolimowski, H., 1994. *The participatory mind.* London: Arkana.

Snyder, G., 1990. *The practice of the wild.* New York: North Point Press.

Torbert, W.R., 1991. *The power of balance: transforming self, society, and scientific inquiry.* Newbury Park: Sage Publications.

Torbert, W. R., 2004. *Action inquiry: the secret of timely and transforming leadership.* San Francisco, CA: Berrett-Koehler Publishers.

Torbert, W. R., 2014. Mathematical intuitions: underlying the meta-paradigm of science named 'collaborative developmental action inquiry'. Unpublished paper.

Part II

Building (tentative) affinities

5 How we nose

Timothy Hodgetts and Hester

Tim's diary: calibrating eyes and noses

The rest of the country has been gripped by a series of winter storms and their stifling snows, but west Cornwall has remained relatively mild – until today. Hester and I begin each morning with a walk up to the lighthouse and the cliffs beyond. We walk bound by a lead, to keep us from chasing rabbits and choughs over the edge (and contributing to the coastguard's alarming statistics about human-dog cliff rescues). Halfway along our route, the path passes through a gate in an old tussocky wall. Each day as we approach, Hester becomes increasingly excited. Her nose is closer to the ground, her sniffs more frequent and vigorous, bent legs lowering her torso so that she seems to dance, crab-like, to and fro. When we pass through the gate, this moment of excitement seems to ebb, and we walk on. Each day, a similar performance. And then this morning we join the rest of the country with our own fleeting mat of sticking snow. Today, Hester's movements inform me, is more exciting than normal. There is more sniffing, more chasing of scents. And when we reach the gate through the wall, I can for once see what Hester smells each and every morning. For whilst in the fields there are sparse animal tracks here and there, through the gate there is a veritable melee of recent traces, excursions large and small. For her, the snow seems to heighten her senses, or at least her excitement; for me, the snow makes scent-lines visible, and makes sense of her dancing.

This chapter is about '*more-than-human participatory research*' in wildlife conservation that involves dogs and humans, and perhaps some pine martens as well. Following the weave of this edited collection, literatures from the diverse fields of participatory research (PR) and more-than-human research (MtHR) are brought into conversation in a productive friction that moves from noting their shared antecedents towards identifying points of generative tension – in this case, literatures specific to wildlife conservation. We consider how the various actions of dogs and martens and humans might be re-imagined as *more-than-human participatory* practices – and what this conceptual cross-stitching might imply for both: (i) the ontology, and (ii) the ethics of research that claims to be participatory with a more-than-human sensibility.

The 'we' in this case comprises one human and one dog. The human did the writing, despite the dog's attempts to contribute by pawing the keyboard and barking suggestions (mostly, the human thinks, relating to the possibility of a game of ball). But we both played distinctive (albeit intersecting) roles in producing particular knowledges. Through a series of auto-ethnographic encounters, we trace how our multi-species attunements produced knowledge within (and for) practices of wildlife conservation; or, put differently, we narrate how the human and dog went surveying for (endangered) pine martens. As we will discuss, taking more-than-human actors seriously as participants in research (rather than as objects, or as methodological tools) has complex ethical implications – and part of our route towards addressing some of these is by unsettling authorial conventions (following mrs kinpaisby 2008).

As well as a dog and a human, this chapter also concerns some elusive pine martens that perhaps live in rural west Wales and that are drawn together with the human and the dog in a specific set of conservation practices. Viewed from the perspective of more-than-human research (particularly as inspired by science studies and/or assemblage theories), *all* of these mammals (martens, dog and human) would be considered as actors who generate a particular but contingent set of social practices – that is, they are all (in a sense) ‘participants’ in assembling what Bruno Latour (the sociologist of science) would term ‘the social’ (2005). By contrast, in participatory research approaches, the notion of ‘participant’ is usually reserved for human agents. It is exactly this question of how to conceptualise ‘participation’ in a more-than-human context that we aim to illuminate through attending to our specific examples of conservation practices; and the decision to not further extend authorial status to the martens hints towards part of the settlement.

Wildlife conservation and participation

Herein, the broad context is provided by practices of wildlife conservation, in which various forms of non-expert and nonhuman involvement each have long histories, albeit differently enacted and understood. In particular, wildlife conservationists frequently rely on multiple human participants to enact various conservation practices (Adams 2004). In the UK, often the starting point for conservation practice is establishing the presence or absence of a particular species (Hinchliffe 2008); and the sheer number of species of concern has necessitated combining diverse sources of information generated by human specialists, professionals, volunteers, lay naturalists and ‘local’ peoples alike (Ellis and Waterton 2004). Yet the extent to which such practices can be labelled as ‘participatory’ can differ markedly.

Practices at the intersection of conservation and development have been a notable concern in PR literatures addressing the relative depth of ‘participation’ in research and politics (Pretty and Shah 1997). Much debate focuses on the extent to which participation actually manages to fulfil (or not) its promise of a meaningful opening of the practices and processes of problem-framing and decision-making to multiple participants to enact social change (Cooke and Kothari 2001, Eubanks 2009). Merely using human volunteers to collect data according to frameworks designed by conservation scientists (as above) would not rate highly in the typologies of participation developed about conservation and development (see Pretty 1995, reproduced in Pretty and Shah 1997, p. 54; see later discussion). Some PR approaches would go further; for example, the participatory geographers Rachel Pain and Peter Francis suggest (albeit in the different context of young people, crime and exclusion) that, '[t]he term "participatory" should be avoided when the primary intention is traditional "extractive" research for the purposes of gathering information' (2003, p. 53).

Within more-than-human research, the focus has been somewhat different in its approach to the extent to which research (including that into wildlife conservation) can be considered and enacted as a more-than-human achievement (Whatmore 2002, Bear and Eden 2011, Lorimer 2015). Influenced by some of the same (and some additional) feminist, science studies and anthropological literatures as PR, such work also draws on a posthumanist sensibility (Wolfe 2010) through which to emphasise the agency of nonhuman actors in a profusion of situations. Of particular relevance to the present chapter, much attention has been paid to the various ways in which human actors can attune to and empathise with nonhuman animals (Despret 2004, 2013, Johnston 2008), often drawing on frameworks of multi-species ethnography (Kirksey and Helmreich 2010), more-than-human ethnography (Barua 2014) or beyond-human anthropology (Ingold 2013) to do so. These attunements enact forms of communication across species boundaries (Kohn 2013, Hodgetts and Lorimer 2015) that might allow marginalised nonhuman voices to be heard alongside (perhaps marginalised) human ones. This, therefore, is a point of both meeting and departure between PR and MtHR, wherein a shared commitment to hearing the unheard is suggestive of different paths based on the types of actors involved.

In the following discussion, we[1] suggest two particular lines of thought suggested by exploring this meeting-place of theoretical approaches, in the context of conservation practices. The first relates to the ontology of participation – i.e. what might constitute participation in *more-than-human participatory* research. Our aim is to foreground some of the ways in which nonhumans might be considered as 'partners' in participatory research, going beyond the by now familiar materialist claim that all manner of nonhuman actors have agency (see, for example, Latour 2005). The second picks up on the ethical implications thus engendered, particularly as regards consent, mutual benefits and recognition. Our argument is that combining the insights of both fields helpfully illuminates the ethical stakes involved when nonhuman actors participate in human-led research on more level, if never identical, terms.

Humandog surveys for pine martens

Pine martens are a cat-sized species of omnivorous mustelid, bigger cousins of stoats, weasels and polecats. They have large ears, long tails, and a bib of light fur on their chests; imagine something a bit like a juvenile fox crossed with an otter and you are on the right lines (a similarity which can confuse sighting records).

Preferentially, they live in the forest canopy, making dens in the fissures of old oaks, safe from the predatory fox below and goshawk above. Centuries ago, these martens were widespread across the UK, but a long history of habitat loss, hunting for their pelts, and finally persecution by gamekeepers led to crashing populations (Yalden 1999). In Wales, they are presently thought to either exist in very low numbers, or to have gone locally extinct. Ongoing conservation projects are working to assess the size and location of any remaining populations, whilst also preparing for a 're-stocking' (Vincent Wildlife Trust 2014).

Searching for these elusive mammals involves various tried and tested ecological survey methodologies, including the use of hair tubes, camera traps and scat surveys. The latter are often preferred for a number of reasons: (i) they are cost-effective in that they require little in the way of equipment, can be conducted over large areas and thus inform the subsequent placement of more expensive cameras; (ii) scat can be sent to the laboratory to be DNA tested for confirmation of species type; and (iii) they are ideal for human volunteers (noting the earlier discussion about the extent to which this might be considered as 'participation') as the techniques are easy to communicate and learn.

Even so, surveying for scat is difficult. Pine marten scat tends to have a particular morphology; but shape can be cryptic, affected by the time of year (and thus diet), weather, size of animal and a number of other variables. It can also be hard to locate; and even when found, the visual similarity of a sample to the scat manual is not always a solid guide to species-identity. Marten scat also has a distinctive smell: It is 'sweet', compared to the rankness of fox excrement – a quality that led to the alternative name 'sweetmart' for this mammal in some areas (Birks and Messenger 2010). Yet human noses (unlike those of their sometime canine companions) are not always sensitive or reliable enough for positive identification. As a result, some conservationists argue that scat surveys are more effective when human surveyors are accompanied by canine assistants (pers. comm., anonymised, May 2013).

What follows is an excerpt from an account of how we learned to scat-survey for pine martens. We were novices together, developing a shared reading of the woodlands we surveyed. Part of the process was learning from both the human and dog 'experts': human instructions on how to identify scat, what procedures to follow when collecting samples for the laboratory; and canine demonstrations of surveying in the woods. The equipment was easy to learn: a pack of ice-lolly sticks with which to lift poo to the nose for a sniff, a pack of re-sealable plastic sandwich bags in which to deposit the sample, a clipboard, a map, a GPS device and a pen. The aim was to use Hester's olfactory sensibilities combined with Tim's sight to locate possible scats, although differential scat identification based on canine scent indication, which would have involved prolonged and specific training for both of us, was not attempted.

***Tim's diary**: It's midsummer, West Wales, near Rhandirmwyn. We are bound once again, this time on a long-lead so that I can see her movements. Luna, the professional scat-sniffing Labrador we met a week ago, works off the lead; Luna is trained to locate scat and 'show' it to her human colleague. Hester and I have* *had no such training, so the lead keeps us connected whilst I look where she sniffs. I've let her smell some sample pine marten scat, but being honest I'm not sure she understood my attempts to explain the game. Instead, I know from experience that she will be drawn to 'interesting smells', and her nose will lead my eyes towards them. And it works. My eyes are becoming nosy. As we walk through the woods, my eyes are often drawn to the animal paths tracing through the bilberry; and Hester's to-and-froing performs a more contemporary tracing. Something went this way very recently, her movements tell me. Then her sniffing springs a squirrel, across the branches and up a tree. WOOF! Up this tree! We are no threat, but foxes would be – and to martens as well. Safer up there. Amongst the trees she leads us to a couple of scats that I wouldn't have seen, and to my nose they smell more sweet than foxy – definitely worth bagging for the lab and later analysis. Pine martens, like many mustelids, are a markedly olfactory species: They communicate over time and space through scent – rubbing, glanding and scat marking their lives, their lines.*[2] *Is it any surprise that through gaining nose, I gain a better window on their world?*

More-than-human participation?

Attunement takes time. It is a process or skill that, as the (more-than) human geographer Steve Hinchliffe and colleagues have suggested, needs to be cultivated (Hinchliffe *et al.* 2005). We thus entered the Welsh woodlands drawing on years of experiences with each other, learning to respond to and improvise with each other and to the unexpected (Laurier *et al.* 2006, Haraway 2008, Brown and Dilley 2012); experiences like the one with which this chapter began, a memorable day in Cornwall in which we both learned to experience landscapes differently.

Woodland *is* experienced differently as a humandog collective. Possible animal tracks, marked by small-footed paths through the bilberry, are given a temporality through the performance and observation of intense sniffing: Tracks recently made provoke more scrutiny than those long-abandoned. Invisible paths across uniform forest roads are made visible by the same means. Eyes become nosey. At one level this is simply about translation: As olfactory clues, whilst not experienced directly or in any of their complexity by the human part of the collective, are instead made available at an 'indexical' level. The concept of 'indexicality' comes from bio-semiotics, and in particular from anthropologist Eduardo Kohn's re-working of Charles Peirce's semiotic theory (Kohn 2013). Therein, unlike abstract 'symbols' or convention-based 'icons', 'indices' instead refer to communicative signs that retain material links to their cause – animal tracks, for example. Whereas symbols and icons largely remain the preserve of humans, 'indexical communication is readily accessible to nonhuman species, including dogs' (Mancini *et al.* 2012, p. 145). Operating as 'index', the animal track enlivened through Hester's sniffing thus has an overlapping, if not identical, meaning for both of us. Whilst the specificities of aromas remain a canine experience, their intensities are bodily translated.

But this translation is not only a representational achievement. Attuning does not rely only on cognitive interpretation of bodily practices, and instead can be theorised as a kind of '*embodied empathy*' – a term coined by the philosopher of science Vinciane Despret to describe the

> feeling/seeing/thinking bodies that undo and redo each other, reciprocally though not symmetrically, as partial perspectives that attune themselves to each other . . . empathy is not experiencing with one's own body what the other experiences, but rather creating the possibilities of an embodied communication.
>
> (2013, p. 51)

This notion of cross-species embodied communication informing much contemporary more-than-human research resonates with a similar riff that is rare, but not entirely absent, in PR literatures – Peter Reason, for example, suggests that, 'the place of humans in the web of life is as embodied participants, 'living as part of the whole'. From this perspective we can begin to articulate a participative worldview' (2005, p. 37). For Despret (2013), embodied empathy can be enacted in various ways as a tool, to attune and respond to, communicate and correspond with nonhuman animals. It is as much 'engaged' as it is 'detached', to use the anthropologist Matei Candea's understanding of both terms not as 'polar opposite' subjective or objective epistemologies, but instead as 'aspects of relationship[s]' (2010, pp. 241, 254). Thus, the translations that occur between human and dog are felt as much as they are interpreted. The dog crouches; the human's muscles tense. The human sees movement and holds still, looking; the dog pricks her ears, tilts her head, and again.

Canine assistance in scat surveys thus works on (at least) two levels. First, dogs are more effective at locating and identifying scat than humans. Yet a more-than-human understanding of such surveying practices goes beyond the instrumental logics that conceptualise dogs simply as a kind of extra-sensory appendage or tool, to highlight the processes of attunement that allow affective communication and negotiation to occur. Thus, second, through a more-than-human lens a surveying dog might also be construed as an active 'participant' in this research practice, inasmuch as the human and dog viscerally collaborate on direction and detection through embodied forms of communication.

Crossing literatures to PR approaches focussed on conservation practices, specifically Jules Pretty's influential typology of participation in conservation programmes, it is tempting to frame Hester's involvement in terms of a shift from '*Participation for material incentives*', and '*Functional participation*' towards '*Interactive participation*' – albeit doing so would involve significantly stretching these concepts (Pretty 1995, cited in Pretty and Shah 1997, p. 54). These terms represent the fourth, fifth and sixth levels of participation respectively: 'material incentives' are fairly self-explanatory; 'functional participation' involves utilising participants as 'means' towards pre-determined goals; and 'interactive participation' describes situations where 'people' (although in this case we might want to widen the category to nonhumans) participate in 'joint analysis, development of action plans' and the like so as to 'take control over local decisions', and thus, 'have a stake in maintaining structures or practices' (Pretty and Shah 1997, p. 54). Although the former categories (material incentives and functional participation) retain a clear resonance with certain aspects of our relationship, they also seem inadequate from a more-than-human perspective; the latter category (interactive) is perhaps closer to capturing the generative and creative roles played by Hester in a series of processes that relied on both of us to continue.

Staying with this typology, the seventh (and last) level of participation Pretty identifies relates to 'self-mobilization', and describes participatory practices that share control of the design as well as the conduct of research. Indeed, for many in PR fields, de-centred and democratised control of research design is the most important criterion for effective, authentic participatory research. There is a sense in which Hester's interest in chasing scents in the woodland might be considered a form of self-mobilisation; but to do so misses the way in which both of us were enrolled in a broader research agenda where the overall design remains in the hands of conservation scientists – reflecting the model of participation in much UK conservation discussed earlier. Indeed, considering this seventh level of participation shifts the axis of analysis, from asking about power distributions between Hester and Tim in our interactions, and instead to asking how we were both enrolled in a wider piece of research as more or less empowered participants.

It is tempting to take this conversation yet further, to stretch the concepts just a little more. Whilst we were both surveying for pine martens as part of a broader piece of ecological research, the design of which neither of us participated in, the particular structure of this research offered opportunities for flexibility in operation. Unlike ecological surveys designed to measure the abundance or distribution of a mammal species, in which factors like surveying method are strictly controlled so as to ensure comparability, we were instead part of a project with a different aim: simply to find any sign of martens that we could, by any means available. This freed us to innovate in ways not dissimilar from some forms of PR. As PR practitioner Jenny Pearce has observed, 'methodologies which sincerely build research processes with practitioners and activists lose considerable control. The creativity lies in the unexpected and contingent' (2008, p. 1). Thus, just as those running the project ceded certain kinds of control to us, we also learned to respond to each other's movements and actions in Welsh woodlands, we developed a form of participation *with each other* more akin to a 'partnering' – a concept that has had a degree of traction in describing humandog relationships, from guide dog partnerships (Higgin 2012) to agility partnerships (Haraway 2008) and many others besides. The notion of 'partnering' also has a distinct and important history in PR; indeed, 'partnership' denotes one of the rungs of Sherry Arnstein's classic 'Ladder of Citizen Participation', in which it is described as a type of participation that enables citizens to 'negotiate and engage in trade-offs with traditional powerholders' (1969, p. 217). However, the MtHR-inspired geographers Emma Roe and Beth Greenhough's notion of 'experimental partnering' is closer to what we have in mind here. They introduce the term to

> define an interpretative approach that is attentive to how practice can illuminate the improvisatory or unstable temporary alignments that underlie some habits . . . [and] that draws attention to human-nonhuman relations and assemblages, fostering new apprehensions of how these more than social relations modify and interrupt the habitual.
>
> (2014, p. 45)

For these authors, experimental partnering offers a means to become more attuned to the agential capacities of nonhuman agents – but it is perhaps best understood as a multi-directional process, not reserved for human actors. Thus, through this partnering and its contingent outcomes, as we collaborated together on unplanned routes through the trees, we began to share a degree of control over our research practices with respect to both each other, and to the wider project in which we were participating.

However, whilst re-framing our shared participation as '*interactive*' or as a form of '*partnering*' would foreground the very real contribution to the production of particular knowledges made by both canine and human surveyors as they collaborate, it would also mask the very unequal relations of power that the typologies of PR practices have done so much to excavate with respect to people. The choices open to dogs, if they can be construed in such anthropocentric terms (which the human author doubts), are often only open with respect to the hand that feeds. Thus the term 'colleague' as utilised in both the diary entry and the discussion above, serves to simultaneously validate the crucial agential role played by dogs, whilst at the same time masking very real unequal relations of power between human 'handler' and dog. Indeed, this is perhaps one of the key areas of friction between critical PR and MtHR approaches (as diverse as *both* sets of literature are), inasmuch as the former emphasises uncovering hidden forms of repression in participatory schemes (Cooke and Kothari 2001), whereas the latter instead amplifies the agency of discursively (and often materially) repressed nonhuman actors – especially nonhuman animals. But it is also worth noting here the post-structuralist critique of all forms of participatory research – that even when characterised by 'deep' or 'strong' participation, it still 'constitutes a form of power that has dominating effects' (Kesby 2005, p. 2038, see also Cooke and Kothari 2001). If power hierarchies are in some sense inevitable in all forms of (participatory) research, the onus on participatory researchers is nevertheless to seek to foreground and dissipate these hierarchies where possible; as Kesby continues, if 'power cannot be avoided, I suggest that it must be worked with' (2005, p. 2038). The context of nonhuman participation serves to make these wider issues of power more apparent.

Considering partnerships also reminds us of the so far silent partners in our story: the pine martens, whose ghostly mobilities shaped our movements. The conceptual power of Roe and Greenhough's 'experimental partnering' lies in its articulation of 'partners-in' as distinct from 'objects-of' research. Following their logic would articulate the martens as partners alongside the two of us in a conservation assemblage or network of relations. But if it is difficult to persuade Hester to perch with Tim on the steps of a typological ladder from PR literature relating to conservation, it is even harder to manage the conceptual gymnastics required to affix martens in place. The pine martens in our stories do not fit, even messily, into categories of participation based on human (inter)action, and their continuing absence from our survey sites also raises questions about what Kesby (2005) has termed the 'spatial dimensions' involved in participation. Indeed, in many ways, this is where these bodies of literature begin to depart into irresolvable difference: they have travelled too far from each other to remain commensurable. Thus, whilst an MtHR reading of our story would incorporate martens as participating partners in enacting a specific set of conservation practices, and a PR interpretation might limit itself to considering the human volunteers, a *more-than-human participatory* reading is faced with the familiar challenge in animal studies of where to draw the line of inclusion (Wolfe 2010). This is, perhaps, the crux of the whole question raised by considering more-than-human participation: What counts? Any settlement of this question, however contingent, partial or temporary it may be, will have significant ethical implications.

The ethics of more-than-human participation

Re-thinking our practices as a form of more-than-human participation thus has several pressing ethical implications. Here, we want to focus on three areas that are often emphasised in PR literatures, and are inflected in important ways by more-than-human concerns: (1) consent, (2) mutual benefit and (3) recognition. Doing so involves a degree of what geographer Catherine Johnston (2008, p. 643) has termed a 'responsible and informed anthropomorphism'; a conceptual move that is widely endorsed in more-than-human research, but is markedly less widespread in PR literatures. First, questioning the extent to which Hester consented to participate in this research, and considering our conservation practices in the woods of Wales, we suggest that Hester's activities might be construed as a form of 'tacit approval'; she is, in Tim's experience, not shy of expressing her disapproval of a suggested course of action, which usually results in the shoulder-hunch and the statue-stance (or if scared, the underbelly tail). Again, however, such forms of attunement will always remain speculative, and it is important to keep the power inequalities in our relationship in the foreground. Furthermore, it is difficult to trace the ways in which any such consent might be deemed as 'informed'. Second, again focussing on the specificities of the situation, we might point towards forms of mutual benefit. Both Hester and Tim seem to enjoy spending time in the woods together; they are both enthused, a spring in Tim's step, Hester very much a springing spaniel.[3] Thus again, the justifications for such claims are to be found in the account of embodied empathetic attunement offered earlier. Yet the benefits of the *outcomes* of the research (in this case, pine marten distribution maps that feed into conservation policy) – which are a crucial test of the extent of participation in PR literature – seem less obviously shared. And this leads us to the ethics of recognition. Although Hester seems very interested in the location of woodland mammals, as far as Tim can make out, she has very little interest in academic publishing. Yet an important aim in PR is to give voice to those whom Askins and Pain (2011, p. 806) term 'producers of knowledge' and thus to coproduce methods of dissemination – and this is a concern shared in MtHR. Our shared authoring is thus aimed at muting the monotonal effects of a sole human voice, albeit that Hester's consent in this case is impossible to affirm (and this remains problematic).[4]

What, again, of the elusive pine martens in all of this? First, we have no way to judge if martens *consent* to our searching for them; although their elusiveness might give us a strong clue since they seem to actively avoid detection. Second, conservationists justify any intrusion on the grounds of an *assumed benefit* for the conservation of the species – albeit not necessarily the individual, with all the ethical issues this raises. And third, in the context of our research there seems little room to *recognise* the absent martens as 'producers of knowledge' (and thus as 'authors') – there are, for example, no instances of shared practices that we can evidently point towards, albeit that there are obvious relational connections between our movements and theirs. Yet in part this is an artefact of the survey design, with the pine martens figured as 'objects' to be counted. A fully more-than-human approach might well seek to destabilise the architecture of such a project, and call for more prolonged ethological engagements through which to emphasise the martens as 'subjects'. But again, there would be ethical implications arising from increasing engagement – or intrusion – in this way.

Thus, perhaps the lesson here is to draw on both traditions (PR and MtHR) to refigure a form of conservation along more participatory lines. To do so, *recognising* nonhumans – perhaps as 'experimental partners' rather than 'objects' (following Roe and Greenhough; see earlier discussion) – seems the easiest move, yet perhaps also requires the biggest cognitive and discursive shift. Designing research for genuine *mutual benefit* seems promising, although the shift towards the logics of 'ecosystem services' in much contemporary conservation practice – which frames conservation action in terms of its benefit to humanity – could be viewed as a step in the wrong direction. And finally, adjusting conservation to incorporate the ethics of *consent* is perhaps the hardest of all. Indeed, requiring consent would rule out many of the tools of contemporary conservation: It is difficult to presume consent to being radio-collared or PIT tagged when the application requires trapping and tranquilising the creature involved. Indeed, for some in animal studies and cognate fields, this is exactly why such a move towards an ethic of consent might be required.

The ethical moves above might be construed as a thinly veiled effort at liberal extensionism, whereby nonhumans are re-articulated as like-humans and granted rights to informed consent and consideration of their 'interests' accordingly. But that is not our intention; and we align with the well-known critiques against such a move (e.g. Hinchliffe *et al.* 2005, Wolfe 2010). Instead, it represents an attempt to think through the implications of considering participation in more-than-human terms, and thus exploring the ethical ties generated through multi-species participatory practices and forms of cooperation. Our claim is that conceptualising those involved in conservation practices not as humans or non-humans but as *participants* is conceptually generative of a new kind of ethical sensibility in which questions about consent, benefits and recognition are re-animated. Not a liberal more-than-humanism, but an ethics of more-than-human participation.

Conclusions

In this chapter, we have explored the intersection of participatory research and more-than-human geographical literatures through reflecting on the shared activities of a dog and a human in a particular set of wildlife conservation practices. Applying both lenses simultaneously produces areas of clarity and confusion. Considering the ontology of participation forces recognition of the different ways in which participation has been understood in the respective fields, but also points towards some of the promise that bringing them together might presage. The conservation practices on which we reflected were not instigated as 'participatory research' in a sense that PR scholars would recognise. Instead, our suggestion here is that our activities as conservation surveyors might productively be re-thought *through* a lens that combines more-than-human and participatory research. Such a conversation is possible because PR is best understood less as a set of strictly delineated methodologies (as in Kumar 2002), but instead more as a set of 'orientations' towards research that imagine participation in terms of meaningful community involvement that leads to 'concrete social change' (Eubanks 2009, p. 115). Our suggestion is simply that the ways in which such 'orientations' are imagined could be widened productively, going beyond the humanist focus of existing PR literatures to include nonhuman communities as part of the societies for whom social change is envisaged and pursued; and that doing so helps to highlight several key ethical questions in research relating to consent, mutual benefits and recognition.

The research on which this chapter is based was funded by the E.S.R.C.

Thanks to the editors whose detailed and perceptive suggestions took this chapter into new and exciting terrain for both of us; and to Jamie Lorimer, Henry Buller and Derek McCormack for constructive comments and conversations on the material herein.

Notes

1 Despite the familiar categorisation of conceptual thought as a uniquely human activity in much Western/modernist thought, insisting on 'we' as authors at this stage serves to foreground how any such thought can be understood as a contingent relational achievement in MtHR; and an achievement in this case that relies strongly on Hester's various contributions.

2 There are thus ethical questions raised by removing scats that serve communicative as well as excretive functions, with a logic of 'the good of the species' used as a

justification in conservation practice, and somewhat mitigated by the tiny number of scats found and removed as a proportion of those potentially produced.

3 Albeit this mention of Hester's 'breed' raises questions about consent and the breeding of dogs for particular characteristics; see Giraud and Hollin, in this volume.

4 In a sense, the situation is somewhat analogous to Wittgenstein's infamous lion, only in reverse: Even if Tim could communicate fluently in the olfactory tones, oral hues and bodily gestures used by dogs, he is not sure that Hester would understand his explanation of 'impact'.

References

Adams, W.M., 2004. *Against extinction: the story of conservation.* London: Earthscan.

Arnstein, S.R., 1969. A ladder of citizen participation. *Journal of the American Institute of Planners*, 35 (4), 216–224.

Askins, K., and Pain, R., 2011. Contact zones: participation, materiality, and the messiness of interaction. *Environment and Planning D: Society and Space*, 29 (5), 803–821.

Barua, M., 2014. Volatile ecologies: towards a material politics of human–animal relations. *Environment and Planning A*, 46 (6), 1462–1478.

Bear, C., and Eden, S., 2011. Thinking like a fish? Engaging with nonhuman difference through recreational angling. *Environment and Planning D: Society and Space*, 29 (2), 336–352.

Birks, J., Messenger, J., and Vincent Wildlife Trust, 2010. *Evidence of pine martens in England and Wales, 1996–2007: analysis of reported sightings and foundations for the future*. Ledbury, Herefordshire: Vincent Wildlife Trust.

Brown, K., and Dilley, R., 2012. Ways of knowing for 'response-ability' in more-than-human encounters: the role of anticipatory knowledges in outdoor access with dogs. *Area*, 44 (1), 37–45.

Candea, M., 2010. 'I fell in love with Carlos the meerkat': engagement and detachment in human-animal relations. *American Ethnologist*, 37 (2), 241–258.

Cooke, B., and Kothari, U., 2001. *Participation: the new tyranny?* London, New York: Zed Books.

Despret, V., 2004. The body we care for: figures of anthropo-zoo-genesis. *Body & Society*, 10 (2–3), 111–134.

Despret, V., 2013. Responding bodies and partial affinities in human–animal worlds. *Theory, Culture & Society*, 30 (7), 51–76.

Ellis, R., and Waterton, C., 2004. Environmental citizenship in the making: the participation of volunteer naturalists in UK biological recording and biodiversity policy. *Science and Public Policy*, 31 (2), 95–105.

Eubanks, V., 2009. Double-bound: putting the power back into participatory research. *Frontiers: A Journal of Women Studies*, 30 (1), 107–137.

Haraway, D.J., 2008. *When species meet.* Minneapolis, MN: University of Minnesota Press.

Higgin, M., 2012. Being guided by dogs. *In*: L. Birke and J. Hockenhull, eds. *Crossing boundaries: investigating human-animal relationships*. Leiden: Brill, 73–88.

Hinchliffe, S., 2008. Reconstituting nature conservation: towards a careful political ecology. *Geoforum*, 39 (1), 88–97.

Hinchliffe, S., Kearnes, M.B., Degen, M., Whatmore, S., 2005. Urban wild things: a cosmopolitical experiment. *Environment and Planning D: Society and Space*, 23 (5), 643–658.

Hodgetts, T., and Lorimer, J., 2015. Methodologies for animals' geographies: cultures, communication and genomics. *Cultural Geographies*, 22 (2), 285–295.

Ingold, T., 2013. Anthropology beyond humanity. *Suomen Antropologi: Journal of the Finnish Anthropological Society*, 38 (3), 5–23.

Johnston, C., 2008. Beyond the clearing: towards a dwelt animal geography. *Progress in Human Geography*, 32 (5), 633–649.

Kesby, M., 2005. Retheorizing empowerment-through-participation as a performance in space: beyond tyranny to transformation. *Signs*, 30 (4), 2037–2065.

Kirksey, S. E., and Helmreich, S., 2010. The emergence of multispecies ethnography. *Cultural Anthropology*, 25 (4), 545–576.

Kohn, E., 2013. *How forests think: toward an anthropology beyond the human*. Berkeley, CA: University of California Press.

Kumar, S., 2002. *Methods for community participation: a complete guide for practitioners*. London: ITDG.

Latour, B., 2005. *Reassembling the social: an introduction to actor-network-theory*. Oxford: Oxford University Press.

Laurier, E., Maze, R., and Lundin, J., 2006. Putting the dog back in the park: animal and human mind-in-action. *Mind, Culture, and Activity*, 13 (1), 2–24.

Lorimer, J., 2015. *Wildlife in the anthropocene: conservation after nature*. Minneapolis, MN: University of Minnesota Press.

Mancini, C., van der Linden, J., Bryan, J., Stuart, A., 2012. Exploring interspecies sense-making: dog tracking semiotics and multispecies ethnography. *In: Proceedings of the 2012 ACM conference on ubiquitous computing, 5–8 September, Pittsburgh, PA*. New York: ACM, 143–152.

mrs kinpaisby., 2008. Taking stock of participatory geographies: envisioning the communiversity. *Transactions of the Institute of British Geographers*, 33 (3), 292–299.

Pain, R., and Francis, P., 2003. Reflections on participatory research. *Area*, 35 (1), 46–54.

Pearce, J., 2008. 'We make progress because we are lost': critical reflections on co- producing knowledge as a methodology for researching non-governmental public action. *NGPA Working Paper Series*. London: London School of Economics.

Pretty, J. N., and Shah, P., 1997. Making soil and water conservation sustainable: from coercion and control to partnerships and participation. *Land Degradation & Development*, 8 (1), 39–58.

Reason, P., 2005. Living as part of the whole: the implications of participation. *Journal of Curriculum and Pedagogy*, 2 (2), 35–41.

Roe, E., and Greenhough, B., 2014. Experimental partnering: interpreting improvisatory habits in the research field. *International Journal of Social Research Methodology*, 17 (1), 45–57.

Vincent Wildlife Trust, 2014. *Pine marten reinforcement feasibility study* [online]. Ledbury, Herefordshire: Vincent Wildlife Trust. Available from: http://www.vwt.org.uk/ [Accessed 7 April 2014].

Whatmore, S., 2002. *Hybrid geographies: natures, cultures, spaces*. London, Thousand Oaks, CA: Sage.

Wolfe, C., 2010. *What is posthumanism?* Minneapolis, MN: University of Minnesota Press.

Yalden, D. W., 1999. *The history of British mammals*. London: T & AD Poyser.

6 An apprenticeship in plant thinking

Hannah Pitt

As this collection demonstrates, co-production as a democratic learning process, while unbalancing typical power hierarchies and engaging with voices from the margins, has clear parallels with the call for social research to embrace nonhumans. In this chapter, I will consider what this means for research with more-than-human communities, placing the tradition of participatory action research in dialogue with attempts to know plants. This conversation seeks a way of thinking through co-production which may redress plants' marginalisation in Western thought (Hall 2011, Marder 2013). Second, it reflects on how 'planty knowledge' challenges participatory research in general.

I suggest the term 'planty knowledge', derived from the notion of plantiness introduced by Head et al. (2012, p. 26), as a way to talk about a loose group of nonhumans which seem to be similar, but which do not neatly fit within a discrete category recognised by biological science or lay understandings. Their deliberately loose word denotes 'an assemblage of the shared differences of plants from other beings' (2012, p. 27), such as the capacity to photosynthesise and having cell walls largely comprising cellulose. Speaking in terms of plantiness seeks to avoid implying that all plants are the same, whilst drawing attention to their particular modes of agency which can otherwise be overlooked or mis-represented (Head *et al.* 2012, p. 29). Planty knowledge then refers to a combination of what humans learn about plantiness, and that which plants themselves understand or sense of the world.

Plants have long been relegated to the lower regions of social hierarchies for being assumed too mute and passive to have anything intelligent to share. This neglect is hugely out of synch with their significance to society and overlooks how intimately entangled they have always been with human lives (Brice 2014, Head et al. 2014). Social scientists countering this neglect have been stymied by a lack of methodologies for including plants as active agents, rather than through representation by human ventriloquists (Head and Atchison 2009, Head et al. 2014). I have previously explored how ethnographic techniques can help researchers to get closer to plants, particularly when guided by experts who grow and tend them (Pitt 2015). I argued that apprenticeships with these human experts offer an important methodology for working with plants in ways that recognise their agency. But what might it mean to regard plants themselves as experts, and can participatory research involve them as knowledgeable participants? To explore this I return to the notion of research as apprenticeship, in which expert guides show what they know of the world, just as novices are inducted into social life (Ingold 2000, Pink 2009). I bring together participatory research and apprenticeship in pursuit of ways for plants to play an active role in co-producing knowledge. Co-produced research meaningfully includes the community in its design and practice through inclusive learning practices which happen with rather than about them (Durose et al. 2011). Co-producing planty knowledge therefore requires work *with* plants in the pursuit of learning to further their needs. A dominant human tendency to neglect plantiness has ecologically destructive effects, so correcting this way of thinking is a step to promote more ecologically sensitive ways of living (Hall 2011). For me, the goal of research as more-than-human co-production is to challenge inequality and injustice which has disempowered and harmed nonhumans. These goals parallel those of participatory action research in the context of marginalised people and communities, which seeks their empowerment; hence I look to this particular tradition to inspire action towards more-than-human communities, which are just for all kinds of beings.

Participatory action offers a focus on seeking positive change and empowerment through research, whilst apprenticeship brings methodological tools to involve plants and planty knowledge in the learning community. A dialogue between these two approaches makes sense because – as I detail later – they speak a similar language regarding the meanings of knowledge, learning and expertise. I bring them together by thinking of the process of doing, and learning to do, research as an apprenticeship. But it is important to note that participatory action research and apprenticeship are associated with divergent notions of community which could be a source of tension. Apprenticeship is said to centre on and form communities united through shared practice and learning (Lave and Wenger 1991); these do not necessarily have any unifying moral vision or common sense of what is good, which some see as an essential component of community (Bauman 2001, Lewis 2006). In contrast, participatory action research seeks positive change, so the community involved has to negotiate which form(s) of good to aim for, developing some common understanding of what is right. Communities of practice united by a shared history of learning (Lave and Wenger 1991) can be seen as oriented to the past, whilst participatory action research communities are oriented towards a better future (Wicks et al. 2008). These distinct orientations have implications for doing research, particularly the question of how communities united by a shared history can transition to become ones shaping a shared future(s). As I will discuss, this transition is complicated further when plants are included in a research community.

Research as apprenticeship creates useful openings for more-than-humans as active participants (Pitt 2015). This 'education of attention' involves experts, such as gardeners, guiding us towards plants by fine-tuning perception towards what they show. But it remains difficult to go beyond what people say and show; disrupting research's conventional power hierarchies requires that plants also be treated as expert guides (Pitt 2015). How might this occur, particularly as the otherness of plantiness remains challenging? I will reflect upon this problem to explore how any participatory research might benefit the communities involved. Along the way I will outline the nature of apprenticeship with more-than-human communities, then consider what we might learn from plants as experts in terms of what philosopher Michael Marder calls 'vegetal thinking'. The final section weaves these strands together with characteristics of participatory action research in order suggest where learning with nonhumans might take us and what barriers may be encountered *en route*.

Co-production as apprenticeship

When seeking to co-produce knowledge with nonhumans I suggest that regarding the process as an apprenticeship is helpful for conceiving the role of the researcher and the skills he or she requires. This is because of several connections between apprenticeship and participatory action research. First, someone learning to do action research does so through a kind of apprenticeship which emphasises skills developed through practical action. Learning happens through engaging in concrete praxis in the world, an education less top-down than typically associated with universities (Levin 2008). A novice action researcher is guided by a mentor who facilitates engagement with practical problems then acts as a critical listener rather than an expert instructor (Levin 2008). Second, groups researching in this way are characterised as communities of practice (Friedman 2001, Reason and Bradbury 2008), the specific form of community that pre-eminent theorists Jean Lave and Etienne Wenger (1991) associated with apprenticeship. These are communities brought together around a common craft, formed through shared learning and a common goal of gaining know-how (Wenger 1998). Similarly, research communities are brought together through the shared practice of learning and a common language developed through this process (Friedman 2001). This suggests that if apprenticeship is to be a basis for more-than-human participatory research, then the first step is to welcome nonhumans into communities of practice.

Lave and Wenger's work on apprenticeship (1991), which informs anthropologist Tim Ingold's understanding of learning, is the foundation for my more-than-human methodology. They present learning as a situated activity, a social process amongst communities of practitioners. Apprentices master skills and knowledge through working with experts, practicing with them in an improvisational process which unfolds through engagement (Lave and Wenger 1991, p. 93). Learning is a process of negotiating and renegotiating meaning, as doing constantly interacts with understanding (Lave and Wenger 1991, p. 51). Novices begin at the periphery of practice, start assisting with small tasks and gradually work towards the centre of the action until they become full members of a community of practice.

Applying this to social research, the researcher can be understood as an apprentice learning by involvement in a community's activity and relations in a thoroughly social process which changes both the community and the researcher–apprentice. This highlights a third connection between apprenticeship and participatory research by mirroring the latter's aspiration for knowledge hierarchies to be disrupted, with the researcher participating as a student rather than an expert (mrs kinpaisby 2008). Further, a master or expert may offer access to a community and hence a route to membership, but she or he is not at the centre of learning or the only source of knowledge; apprentices learn from many, especially their peers (Lave and Wenger 1991, p. 94). A fourth point of connection is that, as with participatory research, the goal of apprenticeship is not theoretical knowledge or abstractions, but practical knowledge which enables action in the world (Lave and Wenger 1991, p. 49). Participatory researchers desire the transformation of participants and their worlds (Kesby 2007, Kindon *et al.* 2007), which also results from apprenticeship. Both the learner and the community change for 'identity, knowing and social membership entail one another' (Lave and Wenger 1991, p. 53). But collectives enacting participatory action research seek a distinctly political form of change with an emphasis on gaining power (Kindon et al. 2007, Reason and Bradbury 2008). Perhaps forming a community of practice is a first stage for research as co-production, necessary to establish relations which provide a basis for the tricky work of challenging injustice. I return to this below, but first consider how communities of practice might include nonhumans.

Lave and Wenger's vision of learning as a social process seems to offer routes in for nonhumans, and could help conceptualise how more-than-human geographers have learnt through immersion in human–plant communities (Hitchings and Jones 2004, Head et al. 2012, Brice 2014). Apprentice–researchers learn planty knowledge by becoming involved in relations with them, participating in their communities and discovering them through interactions. Because apprenticeship is learning through participation and the knowledge it helps to develop is practical rather than abstract (Lave and Wenger 1991, p. 47), it would seem this apprenticeship method could be accessible to nonhuman participants. For example, this method supports non-verbal learning as apprentices do not learn from being told what they need to know but through doing. Talk may feature but is not essential and is not in the abstract form of telling about practice, but rather as a conversation to help explain the doing (Wenger 1998, p. 110). This chimes with participatory researchers' recognition of the value of learning through non-verbal communication (Kindon et al. 2007), and is more open to contributions from nonhuman participants who relate without exchanging words (Buller 2014). Learning proceeds through practice with knowledge being the ability to actively engage in the world (Wenger 1998). Such a focus on doing and living emphasises the material rather than the cognitive, circumventing typical grounds for human exceptionalism and methodological neglect of nonhumans (Whatmore 2006). However, Wenger explicitly precludes nonhumans from learning and communities of practice (1998, p. 138). Although one gains membership through practice, he argues that it is essential that community members share a history of learning, which for him requires negotiation of meaning through communication. Wenger claims that those who presumably cannot experience meaning – a flower, a computer – cannot learn, and so cannot join a community of practice (1998, p. 138).

For Wenger, plants lack capacities for knowledgeable participation in a community of learning. But his exclusively human account of meaning seems at odds with a vision of knowledge as practical, and overlooks potential communication amongst nonhumans, and between humans and nonhumans. Conversely, Ingold's vision of learning as a social process suggests we *can* learn within more-than-human communities. Like Wenger, he approaches learning as a practical, social process of engaging with the world, an improvisational experience of doing things with others (2000, 2013). But for Ingold, the apprentice immerses him or herself in social worlds including birds, fungi and rocks, which might communicate in forms other than mental representations or words. In the case of apprentice hunters, learning involves the nonhuman environment as a novice becomes skilled in perceiving what the earth and plants show (2000, p. 37). All beings, living and non-living, afford things to each other – fungi for eating, waterholes for washing – hence fungi and waterholes are meaningful. What they mean within a particular community is discovered through interacting, with assistance from a guide who shows how to engage with them (Ingold 2000, pp. 20–22). The social world is not exclusively human because 'stones too have histories, forged in ongoing relations with surroundings that may or may not include human beings and much else besides' (Ingold 2011, p. 31). It is not surprising, given the tendency to regard plants as inactive, lacking autonomy or the capacity to act towards a purpose (Jones and Cloke 2002, Hall 2011, Marder 2013), that plants have been excluded from theories of communities of practice. However if we consider the centuries of deliberate breeding which entangle plant species' histories with human activity (Head et al. 2012), then it is misplaced to exclude them for being unable to share a history of learning when it is arguable that they, in fact, do.

Following Ingold, the communities that apprentices learn with can include plants because meaning is not exclusively human. Meaning is not found just in drawing correspondence between the external world and internal representation, but in the coupling of perception and action: perceiving change and responding in order to meet an intention (Ingold 2011, p. 77). This is not an exclusively human skill, as illustrated in the case of animals (e.g. von Uexküll 1992) and plants (e.g. Trewavas 2005). Plants remember an interaction with a person by growing away from the source of disturbance, for example (Chamovitz 2012, pp. 77–82). Humans are a feature of the environment that they sense and understand to a certain extent, and through interactions meaning is exchanged as actions respond to stimulus. Through sharing a history of learning – sensing and responding to each other – and remembering previous interactions, humans and nonhumans become a community of practice. Ingold's is a vision of mutual learning: 'we learn *from* those *with* whom (or which) we study'; for him, learners are not necessarily human (2013, p. 2). If botanists learn *with* plants (Ingold 2013, p. 2), so might social researchers. As active and knowledgeable contributors to social life, there is potential for nonhumans to participate in research as apprenticeship. Ingold does not detail how to learn with plants, or what they might teach. In the next section, I consider what an apprentice co-producing with plants might learn by regarding them as experts in planty knowledge, then move to more tricky questions of what plants might learn.

Planty expertise

Participatory action researchers seek an insider perspective to experience as lived and interpreted by those living it (Reason and Bradbury 2008). Applying this to plants means working to understand plantiness, those shared characteristics common to the biological group which distinguish them from humans (Head et al. 2012). Knowing plants means learning how they live, think and communicate (Hall 2011, Chamovitz 2012). Botanists who seek to see the world from a plant's perspective have come to regard them as active agents able to sense change and respond with intention (Hall 2011, Ch 7, Chamovitz 2012, Marder 2013, pp. 156–160). Daniel Chamovitz details plant activity that is arguably akin to human senses, explaining how they interpret sensory information then respond or communicate it (2012). He holds back from equating this with intelligence, whereas others do on the grounds that plants sense change and select a response suited to enhancing continued living (Trewavas 2005). This coupling of perception and action is what Ingold defines as meaning making; hence, botanical science can be interpreted as supporting the notion that plants make meaning. This opens routes to participation in action research communities, provided this meaning can be exchanged with humans, that is, that we can learn what plants know.

Plants react to changes in the environment and this has enduring effects, a capacity that has been likened to memory. Seeds and developed plants respond variably to stimulus depending on signals they have previously received, showing that they retain a sense of past changes (Trewavas 2003, pp. 7–9, Chamovitz 2012, Ch 6). Plants perceive touch, respond to it and hold a memory of it, just as a person remembers touching a petal. Both are cases of meaning making, so there is potential to share learning, provided humans pay attention to how a plant knows the interaction, through its petals or otherwise.

Marder's (2013) work offers an insightful attempt to redress ignorance of vegetal thinking, meaning the specifics of how plants live. This provides a suitable epistemology for apprenticeship with plants as he does not separate theory and practice: What plants know is what they do. Planty knowledge centres on capacities for growth and nourishment; their being is intentional as stems grow towards the sun and roots reach towards nutrition. Marder highlights how plants arguably have some similarities with humans: He suggests that they communicate through movement or chemicals, not words (2013, p. 75). As we project ideas to understand the world, plants' sensitive roots probe their way towards moisture (2013, p. 27). But he stresses plants' distinctiveness and is careful not to 'offend' them by expecting them to demonstrate human values such as autonomy (2013, p. 55). Whilst encouraging us to recognise plant-like aspects of human life, he urges respect for 'the uniqueness of their existence' (2013, p. 8). Applying this in research might equate to adopting Vinciane Despret's virtue of politeness which she applies to research with animals (2006). She aims to consider what is most meaningful to them and suggests that to discover what this is requires empathetic engagement (Despret 2013). There is more to say about this, but first I want to consider how participatory research with plants might accommodate vegetal thinking.

If plant knowing is plant being, then we learn with plants by being with them, by increasing proximity to them through interactive relationships (Marder 2013, p. 7). This intention parallels the ethnographic endeavour of knowing cultures by participating in their ways of life and relationships (Crang and Cook 2007). For more-than-human research, this means immersion in encounters with plants (Head *et al.* 2012, Brice 2014). The researcher-as-apprentice, seeking to know plants, acts with them, perhaps planting a seed, observing how it responds to heat, improvising a response when leaves start to droop, tracking the action as fruits swell and eventually rot. Marder's philosophical enterprise draws on knowledge from botanists who specialise in such apprenticeship through which they understand plant thinking. For social scientists, a useful route towards an insider perspective on plants is to form a community of learning with these experts who can interpret plants' communication. This approach is vividly demonstrated by Bawaka Country et al., a community of land, researchers and indigenous people expert in more-than-humans in this area of northern Australia, working to know 'the messages that animals, plants, winds send' (2015, p. 275). In their research practice hunters, experts in the area, showed academic researchers how to attend to Bawaka Country's messages through creative engagement with other kinds of beings (2015, p. 10). The research collective suggest that knowing nonhumans starts with respecting them as kin, connected equals who shape each other without domination. The land of Bawaka Country is credited as lead author to reflect that knowledge creation is not a human privilege, that relationships are not hierarchical or centred on humans. In parallel with participatory approaches, their learning aims not to describe the world but to participate with it (Bawaka Country *et al.* 2015).

In my research amongst community gardens in the UK (Pitt 2014), similar perspectives came from permaculture gardeners. Permaculture regards humans and nonhumans as inter-related and inter-dependent, and suggests that ethical living combines care for earth, person and community (de la Bellacasa 2010). These gardeners said that being with plants means letting them do their thing and following how they want to grow. If a flower grows somewhere it has self-seeded that must be the best place for it to grow as it 'knows best', so should be left alone. One gardener instructing me to remove leaves from celeriac plants to benefit their growth suggested the plants would 'tell me' which leaves 'want to come off', so in this context successful gardening means listening to plants. The same gardener said she sees no need to prune fruit trees as they 'work things out' themselves, and grow stronger without human interference. Permaculture displaces human gardeners from a central omniscient position. In this community of practice, people seek to learn with nonhumans and to empathise with what plants might want. An apprenticeship within such communities offers encounters with people and plants to co-produce more-than-human knowledge. This is learning as becoming more skilled in engaging with others (Ingold 2011, p. 223); the end product is not ideas, but the ability to act – or grow – together. Participatory action research's political imperative points to questioning 1) what form this action takes, and 2) for the benefit of who/what. Answering these questions for more-than-human co-production must negotiate plants' otherness.

How plants' silence challenges participation

Marder suggests that to get to know plants requires us to seek encounters with them in which we meet on their terms and respect that these are not fully aligned with ours (2013, p. 7). Plants encountered in Bawaka Country communicate through a language which is not directed towards humans because they are not the centre of things:

> no human beings can really understand all that a rock, a wind or a spear might tell us. These beings have their own ways of being and communicating. This is diversity beyond measure, beyond comprehension.
>
> (Bawaka Country *et al.* 2015, p. 277)

Such more-than-human encounters never offer complete understanding. Often I asked community gardeners why a crop failed or what was wrong with an ailing plant, to be told 'I don't know'; even the most experienced and apparently knowledgeable plants-person struggled to interpret planty being and dying. With a permaculture outlook, such ignorance could be an exciting expression of plants 'doing their thing', but for others, such mystery is a frustrating barrier to growing what humans want to eat. Whilst shared practice can unite humans in a community of learning, this is difficult amongst more-than-humans with diverse ways of practicing. Plants' otherness means there is always something concealed and withdrawn, 'elusive and inexhaustible' meanings (Marder 2013, p. 28). Plants' 'non-thingly' dimensions are those which make them alive, taking them beyond being inanimate things; not easily observed or described, these traits evade us and can make plants seem unreal (Marder 2013, pp. 29–31). The researcher–apprentice might learn with plants through encountering their being, but aspects remain obscure. Humans cannot participate in planty practices like photosynthesis, so the resultant community of practice is one of difference not commonality.

This is not of itself problematic, indeed participatory action researchers are mindful that any community harbours differences which should not be suppressed in pursuit of homogeneity (Gaventa and Cornwall 2008). Ecological communities are necessarily diverse and live through the interaction of organisms with complementary functions – plants produce oxygen which animals breathe in – so researchers might find differences within their communities of practice similarly productive. Where plants' evasive qualities *are* challenging is in relation to the participatory action research's political dimension, the will to achieve change and benefit participants by empowering them to challenge daily injustice (Kesby 2005, Eubanks 2009, Kindon *et al.* 2007). To achieve this, a community of practice must move from a *history* of shared learning, to come together around a shared vision for a better *future*. Here it becomes tricky to maintain a place for plants, and their inability to speak risks researchers speaking on their behalf

(Head et al. 2012, p. 34). Participatory research enacted through relationships of reciprocity rather than paternalism requires that it is not the researcher, but members of the community, who define what is meant by maximising the benefits of involvement (Brydon Miller 2008). But we cannot discuss with plants how they envisage a better future. So researchers seeking to participate with plants face a Catch 22 situation: to assume we know how a plant can benefit from co-production is to reassert the humanist chauvinism which more-than-human approaches seek to overturn. But to expect a plant to participate by expressing its wishes is to see their participation in very human-centric terms which fail to appreciate plantiness. Further, critiques of participatory research suggest that offering benefits of participation is itself paternalistic and can perpetuate power inequalities (Cooke and Kothari 2001).

Can we imagine from a more plant-centred perspective what plants might seek from participatory research? Is it inevitable that in seeking to benefit plants we perpetuate inequality through the tyranny of participation? If there *is* a goal guiding plants' actions, it is most likely 'maximal fitness, usually regarded as the greatest number of viable seeds' (Trewavas 2003, p. 2). To empower a plant can be conceived as enhancing its ability to reproduce, a very different benefit than that which human participants might hope for. But the two coalesce around the idea of flourishing as one way of conceptualising the goal of action research (Reason and Bradbury 2008, p. 4). In planty terms, *flourishing* might suggest allowing plants space to grow and reproduce without always having to serve human purposes (Hall 2011). Permaculture gardeners letting plants 'do their thing' left them to reproduce and grow irrespective or in spite of human benefits; these plants flourished. Co-production directed towards the flourishing of more-than-human communities therefore seeks to allow all the freedom to live out their ways of being. Questioning the form and dynamics of such a process is helpful as a reminder of four tricky issues, equally pertinent to human focused participatory research.

Planty expertise and the tricky challenges of participatory research

Co-production as apprenticeship is learning through practical engagement with others; it proceeds through social relations, and results in a community of practice whose members share a history of learning. Nonhumans participate in these communities but have multiple modes of being which can also be distinct from those of humans; these are communities with difference. Apprenticeship with nonhumans appreciates diverse forms of knowledge including distinct modes of being planty, their speeds, scales and languages. But there are challenges moving from a more-than-human community based on shared history, to one working towards a shared vision for a better future. These four challenges are similarly tricky for participatory research with human communities so they remind us that these are also communities of difference, with power dynamics which researchers act within and on.

1) How much can be known?

The researcher–apprentice works to understand planty knowledge, aware that plants retain some obscurity which humans can never access, so learning with them is always partial (Marder 2013). This inherent unknowability is not uniquely vegetal, research with any community results in partial understanding. But incomplete understanding can be increased as a result of people's will to remain silent by not participating or not speaking of certain things (Dodson *et al.* 2007). The participatory research tradition has largely equated voice with power and sought to hear from those whom society silences (Maguire 2001). But there are various reasons why people might choose to remain silent, including the wish to avoid self-incrimination or to reduce risk of repercussions from speaking out against the more powerful (Dodson *et al.* 2007). The inevitable silence of plants urges us to remember there can be power in declining to speak and in deliberate obfuscation, particularly whilst participation does not always equal empowerment and can imply subordination (Cleaver 2001). In learning with plants, we must remember that we are not authoritative, and so we need to ask what plants have and have not shared with us (Head *et al.* 2014, p. 867). Applying this to human participants encourages respect for any knowledge and experience which they wish to remain hidden, and a reflection on how silence affects what is known. This is humble research accepting uncertainty and difference (Janes 2016), and is dependent on a researcher's ability to cede the will to control.

2) How to agree the goals?

Participatory research is driven by a will to make practical change to maximise benefits for participants (Kindon *et al.* 2007, Manzo and Brightbill 2007). We have seen the complexity of understanding how more-than-human communities can benefit when plants seek outcomes peculiar to us and thrive in unique ways. This highlights that we should never assume what change participants seek because they do not necessarily share our goals. Whilst people might advise on their goals for research encounters, planty participants cannot so humans are likely to determine the change sought. The gardener listening to a celeriac plant cannot be sure that it wants certain leaves removed, as she or he is interpreting wishes conveyed across a language barrier. Another person may 'hear' another message, and another plant may wish to 'say' something wholly different. Those who facilitate research with human communities are aware of the difficulties of ensuring all participants truly benefit (Kesby 2007), and of the need for participants to define what this means (Brydon Miller 2008). Less clear is how best to proceed when participants – like plants – are silent on how they might want to benefit. The question becomes whether to cease research or continue in an inevitably paternalistic mode. Do we assume plants are best served by helping them maximise reproduction? And if this is the goal, how do we negotiate cases when one plant's fitness impedes the development of a neighbouring one competing for the same resources? Or when human benefits conflict with those of plants? These dilemmas offer a useful reminder that it is not always clear how research can benefit the communities involved, and that participants may not all seek the same future (Gaventa and Cornwall 2008).

A route through these dilemmas might be to consider the spatio-temporal scale of benefits; we might not be sure that plants participating with us here and now benefit from participation, but perhaps plantiness more broadly will participate, and in the future. For Bawaka Country, more-than-human research entails understanding ourselves differently by enacting an ethic of care based on a relational view of the world (2015). Similarly, Marder describes how realising plants' lack of autonomy and their dependent relationship with their environment can challenge humans' tendency to objectify the world and extract themselves from it (2013, p. 71). Vegetal thinking leads us to understand the world as 'shared divisibility' in which lives are not self-contained or discrete but inter-dependent, so we are more likely to respect nonhuman lives (Marder 2013, see also de la Bellacasa 2010). The result of an apprenticeship with more-than-human communities is what Ingold calls 'transformation of thinking and feeling' (2013, p. 4), a new way of regarding the world which recognises human commonality with plant being, that lives are not self-contained and flourish through interdependence. We can hope that bringing nonhumans into the realm of ethical regard in this way benefits plant lives in the long-term.

3) . . . And for who or what?

Asking what change co-production can achieve leads to inevitable questions of who or what benefits. Again planty being brings to mind tricky issues for participatory action research. An emphasis on consciousness raising can make participatory research overly focused on the individual, so change at the collective level or wider factors in social injustice are neglected (Kesby 2005, Eubanks 2009). But it is difficult to think of plants as autonomous individuals because their being is dispersed and divisible rather than contained within a thing-like object. A plant's life flies off through seed dispersal, carries on when a twig is removed and transplanted as a cutting, hence a plant's identity 'explodes' into many parts with no clear boundary (Marder 2013, pp. 42–43). Plus a plant's ability to achieve maximum fitness is 'indissolubly linked with the local environment' (Trewavas 2003, p. 8). Plant thinking encourages us to regard all lives as inter-dependent. So would 'empowering' an individual plant to maximise seed production necessarily require that its environment is also 'empowered'? As Bawaka Country and permaculturists remind us, no being lives discrete from others, individual good depends on communal good. This prompts a collective perspective on empowerment in which individuals only benefit from research if their wider community is empowered.

Achieving this in practice is unlikely to be easy. With plants a researcher has to evaluate whether the appropriate unit of ethical engagement is the individual plant, its community or species (Atchison and Head 2013). Translating this to work with human communities presents a potential dilemma of whether to disappoint a particular person in order to serve the wider needs of their community. The ethical response to such dilemmas begins with researchers and communities discussing difference and being honest about the degree to which deeply ingrained power imbalances are levelled through participation (Janes 2016). This pragmatism might also include tempering expectations of participatory research: not all communities of learning become communities of empowerment, and not all problems can be solved.

4) Can empowerment be extended across time and space?

A final and related dilemma is whether benefits of action research remain locally confined. Experts in participatory research recognise the difficulty of extending empowerment across space and time to benefit participants beyond the research encounter (Kesby 2007, Kindon *et al.* 2007, p. 24, Askins and Pain 2011). Research with plants faces similar challenges, which can be confounded by plants' dispersed identity and complex time-scales. How should we empower a plant which might not live beyond today, which will produce offspring in a far flung location, or which only survives through connection with soil and air? What if celeriac only flourishes through killing dandelions? At what scale is the benefit of empowerment to be realised: the local plant or its global species? Human communities are similarly connected, multi-scalar and divided, so again plantiness reminds us of the complexity of achieving change through participatory research. Planty being is inseparable from the wider ecosystem. Similarly, the context of an individual research encounter links to wider flows and relations. This provides a useful reminder of the need for a macro perspective on empowerment and change which considers wider forces and power relations shaping participants' experiences.

One way for research to extend empowerment is by encouraging reflexivity. Participatory research's transformational potential lies in researchers and participants reflecting on their learning and their self, taking this forward as an enhanced ability to make change (Kindon et al. 2007, Gaventa and Cornwall 2008, Reason and Bradbury 2008, Durose *et al.* 2011). Such conscientisation is possible when learning with plants but this will be dominated by human reflexivity; plant encounters are one-sided because plants are not aware that they are in relation with a human individual. A plant's experience of the world lacks any personal awareness and 'a brainless plant likely does not worry about its dignity' (Chamovitz 2012, p. 173). Plants have intention but no consciousness, self or identity (Marder 2013), so learning – no matter how participatory – cannot make them more conscious. If there is no plant-self to partake in self-reflection, they are incapable of consciousness raising; their sense of self and self-awareness cannot change. In this sense, participatory research may always be more transformational for the human apprentice–researcher than for the plant-participant.

We have seen how learning with plants can change people, and that encountering planty being can alter people's view of the world, but this is *our* endeavour, not a plant's. Plants can never be full participants, hence the research can never by truly 'for them' as co-production intends. They are always partially withdrawn from the process, and to allow plants to flourish is to accept a way of being which does not include the will to co-produce. Perhaps other beings also flourish through being left alone, free from pressure to participate. If apprenticeship in planty knowledge is most transformative for the researcher, we should question whether much participatory learning is more important to academics than to participants. We cannot be sure because, as we learn from plants, there is always much we cannot know about others.

These tricky issues lead to two concluding points. First, dialogue between participatory action and more-than-human researchers can continue to exchange on these and other dilemmas to benefit both traditions. Second, co-production with nonhumans is relatively early in its evolution so there remains much to think through, and many mistakes to be made. As for all apprentices, the way for researchers to become more skilled in planty co-production is to keep improvising, and to keep learning by doing with plants.

References

Askins, K., and Pain, R., 2011. Contact zones: participation, materiality, and the messiness of interaction. *Environment and Planning D: Society and Space*, 29 (5), 803–821.

Atchison, J., and Head, L., 2013. Eradicating bodies in invasive plant management. *Environment and Planning D: Society and Space*, 31 (6), 951–968.

Bauman, Z., 2001. *Community. seeking safety in an insecure world.* Cambridge: Polity Press.

Bawaka Country, Wright, S., Suchet-Pearson, S., Lloyd, K., Burarrwanga, L., Ganambarr, R., Ganambarr-Stubbs, M., Ganambarr, B., and Maymuru, D., 2015. Working with and learning from country: decentring human authority. *Cultural Geographies*, 22 (2), 269–283.

Brice, J., 2014. Attending to grape vines: perceptual practices, planty agencies and multiple temporalities in Australian viticulture. *Social and Cultural Geography*, 15 (8), 942–965.

Brydon Miller, M., 2008. Ethics and action research: deepening our commitment to principles of social justice and redefining systems of democratic practice. *In:* P. Reason and H. Bradbury, eds. *The Sage handbook of action research, 2nd Edition.* London: Sage, 199–210.

Buller, H., 2014. Animal geographies I. *Progress in Human Geography*, 38 (2), 635–650.

Chamovitz, D., 2012. *What a plant knows*. Oxford: Oneworld.

Cleaver, F., 2001. Institutions, agency and the limitations of participatory approaches to development. *In:* B. Cooke and U. Kothari, eds. *Participation: the new tyranny.* London: Zed Books, 36–55.

Cooke, B., and Kothari, U., eds., 2001. *Participation: the new tyranny.* London: Zed Books.

Crang, M., and Cook, I., 2007. *Doing ethnographies.* London: Sage.

de la Bellacasa, M., 2010. Ethical doings in naturecultures. *Ethics Policy and Environment*, 13 (2), 151–169.

Despret, V., 2013. Responding bodies and partial affinities in human-animal worlds. *Theory, Culture and Society*, 30 (7–8), 51–76.

Despret, V., 2006. Sheep do have opinions. *In:* B. Latour and P. Weibel, eds. *Making things public: atmospheres of democracy*. Cambridge, MA: M.I.T. Press, 360–370.

Dodson, L., Piatellie, D., and Schmalzbauer, L., 2007. Researching inequality through interpretive collaborations: shifting power and the unspoken contract. *Qualitative Inquiry*, 13 (6), 821–843.

Durose, C., Beebeejaun, Y., Rees, J., Richardson, J., and Richardson, L., 2011. *Towards co-production in research with communities*. Swindon: AHRC Connected Communities.

Eubanks, V., 2009. Double-bound: putting the power back into participatory research. *Frontiers: A Journal of Women Studies*, 30 (1), 107–137.

Friedman, V., 2001. Action science: creating communities of inquiry in communities of practice. *In*: P. Reason and H. Bradbury, eds. *Handbook of action research, 1st Edition*. London: Sage, 159–170.

Gaventa, J., and Cornwall, A., 2008. Power and knowledge. *In*: P. Reason and H. Bradbury, eds. *The Sage handbook of action research, 2nd Edition*. London: Sage, 172–189.

Hall, M., 2011. *Plants as persons*. Albany, NY: SUNY Press.

Head, L., and Atchison, J., 2009. Cultural ecology: emerging human-plant geographies. *Progress in Human Geography*, 33 (2), 236–245.

Head, L., Atchison, J., and Gates, A., 2012. *Ingrained*. London: Ashgate.

Head, L., Atchison, J., Phillips, C., and Buckingham, K., 2014. Vegetal politics: belonging, practices and places. *Social and Cultural Geography*, 15 (8), 861–870.

Hitchings, R., and Jones, V., 2004. Living with plants and the exploration of botanical encounter within human geographic research practice. *Ethics, Policy and Environment*, 7 (1), 3–18.

Ingold, T., 2000. *The perception of the environment: essays in livelihood, dwelling and skill*. London: Routledge.

Ingold, T., 2011. *Being alive*. London: Routledge.

Ingold, T., 2013. *Making*. London: Routledge.

Janes, J., 2016. Democratic encounters? Epistemic privilege, power, and community-based participatory action research. *Action Research*, 14 (1), 72–87.

Jones, O., and Cloke, P., 2002. *Tree cultures: the place of trees and trees in their place*. Oxford: Berg.

Kesby, M., 2005. Retheorizing empowerment-through-participation as a performance in space: beyond tyranny to transformation. *Signs*, 30 (4), 2037–2065.

Kesby, M., 2007. Spatialising participatory approaches: the contribution of geography to a mature debate. *Environment and Planning A*, 39 (12), 2813–2831.

Kindon, S., Pain, R., and Kesby, M., eds., 2007. *Participatory action research approaches and methods: connecting people participation and place*. London: Routledge.

Lave, J., and Wenger, E., 1991. *Situated learning: legitimate peripheral participation*. Cambridge, UK: Cambridge University Press.

Levin, M., 2008. The praxis of educating action researchers. *In:* P. Reason and H. Bradbury, eds. *The Sage handbook of action research, 2nd Edition*. London: Sage, 669–681.

Lewis, T., 2006. Heterogeneous community: beyond new traditionalism. *In:* S. Herbrechter and M. Higgins, eds. *Returning (to) communities: theory, culture and political practice of the communal*. Amsterdam: Rodopi, 55–72.

Maguire, P., 2001. Uneven ground: feminism and action research. *In:* P. Reason and H. Bradbury, eds. *The Sage handbook of action research*. London: Sage, 59–62.

Manzo, L., and Brightbill, N., 2007. Toward a participatory ethics. *In:* S. Kindon, R. Pain, and M. Kesby, eds. *Participatory action research approaches and methods: connecting people, participation and place*. London: Routledge, 33–40.

Marder, M., 2013. *Plant thinking: a philosophy of vegetal life*. New York: Columbia University Press.

mrs kinpaisby, 2008. Taking stock of participatory geographies: envisioning the communiversity. *Transactions of the Institute of British Geographers*, 33 (3), 292–299.
Pink, S., 2009. *Doing sensory ethnography.* London: Sage.
Pitt, H., 2014. *Growing together an ethnography of community gardening as place making.* Thesis (PhD) Cardiff University. Available from: http://orca-mwe.cf.ac.uk/53953/ [Accessed 15 May 2016].
Pitt, H., 2015. On showing and being shown plants – a guide to methods for more than human geography. *Area*, 47 (1), 48–55.
Reason, P., and Bradbury, H., 2008. Introduction. *In:* P. Reason and H. Bradbury, eds. *The Sage handbook of action research, 2nd Edition.* London: Sage, 1–10.
Trewavas, A., 2003. Aspects of plant intelligence. *Annals of Botany*, 92 (1), 1–20.
Trewavas, A., 2005. Green plants as intelligent organisms. *Trends in Plant Science*, 10 (9), 413–419.
von Uexküll, J., 1992. *A foray into the worlds of animals and humans.* Minneapolis, MN: University of Minnesota Press.
Wenger, E., 1998. *Communities of practice.* Cambridge, UK: Cambridge University Press.
Whatmore, S., 2006. Materialist returns: practising cultural geography in and for a more-than-human world. *Cultural Geographies*, 13 (4), 600–609.
Wicks, P., Reason, P., and Bradbury, H., 2008. Living inquiry: personal, political and philosophical groundings for action research practice. *In:* P. Reason and H. Bradbury, eds. *The Sage handbook of action research, 2nd Edition.* London: Sage, 15–30.

7 Imagination and empathy – Eden3

Plein Air

Reiko Goto Collins and Timothy Martin Collins

The problem of discourse and participation

The key question of this chapter is this: Can artwork contribute to the evolution of a tree from a thing to a being of value? Building on over 20 years of work, the Collins & Goto Studio proceeds slowly, immersed in communities of interest and expertise. Working across arts media, digital media, philosophy, science and technology, we explore meaning and interrelationship through experience, discourse, reflection and in-studio practice. Early work was influenced by discourse theory and ethics moderated by ideas about power and agonistic plurality. That work was about changing ideas of the form, function and meaning of nature in a postindustrial context.

In our current work, we are developing a sculptural instrument that supports empathic experience with trees. The work is influenced by Edith Stein's (2002) theory of empathy; published from her doctoral dissertation.[1] This research extends our earlier work to interrelationship with nonhuman others. It contributes to current work in the field of art through attention to empathy with trees in practice. Empathic exchange is enabled through technology-enabled processes; the final form emerges through iterative method. The use of empathic projection and informed imagination raises issues in communication ethics.

In our reading of the participatory research literature, we were struck by the idea of participation as a path to tyranny in the context of international development. Bill Cooke and Uma Kothari describe a tyranny driven by the intentions of facilitators (2001, pp. 1–15): the potential for collective bias towards power and selectivity driving others out of participatory planning. They thus critique the culture of international development and its intentions and methods. This 'problem of tyranny' as intentional culture/institutional maintenance of the status quo, would be keeping with John Friedmann's (1987) thinking about planning. He examines the philosophical roots of allocative, innovative and radical planning in the public interest. He is clear about the cause and effect of intentions and methods, the ethical impact of institutional interests and the practices and relationships that differentiate the maintenance of the status quo from evolutionary change and structural transformation (Friedmann 1987, pp. 51–86). Can development interests meet their own institutional agenda and serve a public participation agenda at the same time? In Cooke's (2001) follow-on chapter on the social and psychological limits of participation he examines theories of collusion and false agreement, groupthink and brainwashing. He understands participation as 'yet another technology used with the Third World without the care and concern that would be expected elsewhere' (Cooke 2001, p. 120). This is a practical realm of powerful interests and conflicted manipulative agendas. John Friedmann suggests that institutions and state interests are ill suited to the emancipatory intentions of radical planning (1987, p. 306). Despite the radical intentions of the participatory planners, the problem of tyranny is more likely located in the institutional drive for development than the participative methodology; although the ethics would be socially revealed in practice.

Our regular reading spans art history, theory and practice. There are three relevant areas of work: related to place, related to environment and related to dialogue. The art critic Rosalyn Deutsche has examined art and spatial politics. She focuses on places where relationships between self and other are constantly settled and unsettled. She uses feminist critique and its relationship to identities formed in public space, to argue for a continuous disruption of interest to prevent the conversion of public space to private interest (Deutsche 1996, pp. 326–327). Miwon Kwon follows with an assessment of community practice and critical social practice in the arts. She concludes with the idea (following Jean-Luc Nancy), 'only a community that questions its own legitimacy is legitimate' (Kwon 2002, pp. 154–155). It helps to consider that the tyranny previously described occurs where participation is part of a 'toolbox' to deliver development agendas and resources. In counterpoint, the artist as a cultural agent is concerned with power and the evolution of social creativity and subjectivity. The artist more often than not works in the interstitial spaces ignored by development interests.

There is a specific history of social and environmental art practices. Much of the early work in the field was about an integration of ideas and materials, a creative inquiry about life on earth. More recently, with greater attention given to environmental change, the practice and critical theory focuses on critical and transformative approaches to issue specific practices. Beginning in the early 1980s, Lucy Lippard (1983, 1997, 2007, 2014) and Suzi Gablik (1984, 1992) began to reveal a new social and environmental context for making art. They shared a critical unease with the art world and outlined theories and described extant practices that integrated art, society and aspects of the environment. Rosalyn Deutsche (1996) theorized the agonistic role of artists working in urban settings and Miwon Kwon (2002) interrogated practices based on lived experience and relationality tied to place. Barbara Matilsky (1992), Jeffrey Kastner and Brian Wallis (1998) are noted for their work describing the environmental art field and its historic relationships to land and earth art. More recently, T.J. Demos, a theorist and curator working on the relationship between art, environment and politics, has challenged traditional criticism by asking cultural interests to think more strategically (2010, 2012, 2013, 2015). The critical view of this work continues to evolve; although somewhat more slowly than the practice.

Another important touchstone is the work of Grant Kester, who writes about the history, theory and practice of social and dialogical aesthetics (2004, 2011).[2] He ascribes value by considering the inter-subjective ethics and indications of empathy that attend creative 'dialogic' arts practices (Kester 2004, pp. 107–115). He also curated an exhibition and edited a catalogue examining the international artists that engage planning and policy (Kester 2005).[3] Artist Jay Koh reflects on his international practice, considering justice-based approaches to subjectivity in 'Art-Led Participative Processes' (2015). Koh is attentive to ethics and competing intentions. He eschews singular value and belief systems as criteria for judgement. Arguing that the facilitating artist occupies a 'decentred position within reciprocal communicative relationships, so as to reduce the power imbalance(s)' (Koh 2015, pp. 161–162) There is significant overlap between Koh, Deutsche and Kwon and the ethical principles and performative methods explored by participative research authors such as Banks *et al.* (2013); the 'methodological modesty' and pragmatism of Eubanks (2009); as well as Askins and Pain's ideas about materiality and an ethics of care (2011). The ethical and empathic issues that were raised by colleagues such as Kester and Koh are part and parcel of our own understanding of the challenges of art practice, as discourse, dialogue and research across the boundaries of species.

The research

In the early part of the new millennium, various long-term ecological research programs were set up by the National Science Foundation in the United States; including forest studies that explored the impact of an increase of carbon dioxide on trees. One of these was the Free Air Carbon Enhancement (FACE) experiment in North Carolina, where ecologists and biologists raised the CO_2 concentration in a specific area of forest, then wired the forest to measure the effect on plant/tree growth. During our visit, outdoors and four stories up in the canopy, we experienced an epiphany of sorts; a sense that we had engaged the essential nature or meaning of the tree; a simple intuitive leap precipitated by a flow of data. While watching the instrument sensor displays, we realized we were 'seeing' the invisible breath and sap flow of a tree. Sap flow was driven by temperature, while photosynthesis (light energy captured in leaves) and transpiration (the loss of water through the leaves) was shaped by available sunlight, localized atmospheric CO_2 (in constant flux) and humidity. This experience led to the development of a project plan using an experimental framework to develop sculptural instruments and technologies that enable exploration and exchange with plants and trees.

As we initiated this project, we examined artwork with trees; projects widely recognized as historically important. In 1978, Alan Sonfist developed a native forest restoration *Time Landscape* in New York City (Sonfist 2007). In this work, the artist challenged ideas about what artists do and what it is that deserves our aesthetic attention. In 1982, Joseph Beuys initiated a symbolic planting of *7000 Oaks* in Kassel Germany (Willisch and Heimberg 2007). At the opening of the seventh Documenta art exhibition in Kassel, the artist placed 7,000 basalt columns in front of the Fridericianum museum, which were to be planted alongside 7,000 trees throughout the city. In 1993, Helen Mayer Harrison and Newton Harrison's completed *Serpentine Lattice* arguing for new policies to support the coastal redwood rainforest of the Pacific Northwest of North America. The project forms an illustrated narrative that initiates sympathetic response. It begins: 'North America's Last Great Temperate Rain Forest is Dying' (Harrison and Harrison 1993, p. 3). Beuys and Sonfist were focused on public aesthetic value and experimental cultural policy and practice related to trees and forests. The Harrisons take a sympathetic approach, which is closer to our own interest in relational aesthetic experience. We also considered recent science and technology informed artwork such as: Chris Chafe and Greg Niemeyer's *Oxygen Flute* (2001),[4] which used sensors and a sound system in an enclosed environment with plants to 'hear' gas exchange in California. David Dunn of New Mexico conducted art-led research on the sound of invasive bark beetles. The composition *The Sound of Light in Trees* (2006),[5] included reflective publication with a scientist. The Collective group 'Active Ingredient' from Nottingham UK collaborated with climate change scientists to develop technology-based work integrating temperature, humidity, carbon dioxide, light, color and sound conditions in artworks such as *Dark Forest* (2009),[6] and *A Conversation with Trees* (2010).[7] More recently, James Bulley and Daniel Jones have produced *Living Symphonies* (2014),[8] presented in various forests in the UK, the authors claim it as a musical composition that 'translates a forest ecosystem' resulting in an ever-changing sound experience. This contemporary work is about technology, sonification, creative expression and musical composition; it engages the artist–viewer relationship. We use some of the same techniques, but artistic authorship is in service to human–nonhuman interrelationship. For us the work begins with empathic intent, which we explain in detail below.

The move towards empathic research

Previous work in Western Pennsylvania *Nine Mile Run* (1997–2000),[9] and *3 Rivers 2nd Nature* (2000–2005),[10] dealt with the restoration of postindustrial ecosystems. Working with scientists, we initiated new knowledge about waterfront ecology and refined key points for public presentation and discussion. Out of this came ideas of strategic knowledge; concepts that had the potential to reshape human perception, experience and normative values – the structure upon which decisions and policies are made. The failure of the industrial economy in the local area had enabled a 30-year natural recovery and produced a community aware of the aesthetic value of restored landscapes. Extant land-use policies and private development interests saw forest cover and functioning ecosystems as an impediment to progress on steep hillsides and riverfronts. Nature could be seen as an agglomeration of subjects needing release or abeyance from the constraints of industrial and postindustrial culture. In response to this, we organized research with communities that wished to consider nature in new ways.

We entitled the new series of work *Eden3* and aimed to consider how empathy could support the development of moral duty through art practice.[11] The work engaged the cycle of carbon/oxygen exchange; attending to the breathing in of one species (trees) and the breathing out of another (humans). Why trees? Trees are the largest most ubiquitous living 'things' in the world. They provide aesthetic, life-enhancing benefit to a myriad of living things. Trees do not have the same bio-chemical senses, the physical mobility or the same kind of agency that humans do. They respond to light, temperature and humidity as well as the chemistries of soil and air. Humans and trees share and shape the environment in distinct and different ways. For the most part, we do not perceive a tree's subtle and quick responses to changes in the environment; this makes them appear to be significantly other to us. Yet some of us have an ability to read the physical state of plants and trees; most recognize life and death over time and seasonal changes. It is possible to sense/see vitality or ill health in plants. Many of us that spend time with plants can recognize complex shadings of well-being linked to available moisture, soil, light, nutrients or predation.

Later, we focus on the developmental tracks of this research: theories of empathy, work with sensors and sound and the form and practice of working toward empathic exchange with trees. The word *empathy* comes from the Greek word *empathis* (*em* + *pathos*) referring to passion, feeling and emotion. In the late 19th century, the theory of empathy became a particular concern within philosophy (Stueber 2014). The word *empathy* translated into English from the German word *Einfühlung*, meaning feeling into. It was not a literal meaning of going inside of the other person, but relying on careful observation and non-verbal communication such as facial expression, eye contact, body gestures and other behaviors and experiences beyond one's intellect. Theodor Lipps, a 19th-century German philosopher, conceptualized this human ability to understand the other as an inner imitation. Phenomenologists such as Edmund Husserl, Max Scheler and Edith Stein expanded on this idea in the early 20th century. Husserl claimed our consciousness is always active and directed toward others and the environment. Phenomenology was a reconsideration of emotions that are not located in discrete bodies and persons, and thus by expression, proposes affect as a phenomenon anterior to the distinction of persons (Burgess 2011, pp. 289–321). Husserl, Scheler and Stein shared a common ground, understanding 'empathy as a kind of act of perceiving, sui generis' (Stein 2002, p. 11). For these theorists, empathy is a perception that is unique, of its own kind. Stein was open to the application of empathy beyond human-to-human relationship; this position informs our work.

The idea of plants and trees as sentient or sensory aware subjects with memory remains controversial. Until recently, mainstream science has been unwilling to consider ideas like sensory perception, communication, memory, agency and knowledge in plants. But there are some cracks in that armor. Biologist Anthony Trevawas has written a series of articles (2003, 2005) that explore 'plant intelligence', arguing that plants are territorial and competitive; forever changing their 'architecture, physiology and phenotype' in the intelligent pursuit of resources for growth and reproduction (2005, p. 413). More recently, Daniel Chamovitz

(2012) argues for awareness (rather than intelligence). He makes a case at the bio-chemical level for specific sensory perceptions, and a form of memory enabling response to changes in the environment. It is important to note this work has vociferous critics, Richard Firn's (2004) response to Trevawas's 2003 paper, makes a point-by-point rebuttal before demanding limitation on anthropocentric description. We would argue this moral constraint against anthropocentrism must be reconfigured as a caution; to intend no harm or overt obfuscation, rather than a line separating humanity and its range of world-views from everything else in the world.

Theories of empathy

> Empathy . . . is the experience of foreign consciousness in general, irrespective of the kind of the experiencing subject, or of the subject whose consciousness is experienced.
>
> (Stein 2002, p. 11)

For Stein, empathy is a practice that can be developed and refined through intimate attention to people and things over time. Following Stein, empathy is an act of perceiving in which we reach out to the other to grasp his/her/its state or condition. It is based on one's emotional and physical experiences. Empathic experience is focused towards something foreign rather than something familiar. It motivates something within that enables different forms of expression than we could know on our own. It adds something to the world that would not otherwise exist. In contrast, Stein understands sympathy as assuming feeling in another based on what we already know about our self and our interests (2002, pp. 14–18). In Stein's theory of empathy, outward expression is a symbol of body and mind relationship. Stein says that 'a sad countenance is the outside of sadness' (2002, p.77). Goto understood that facial expressions can be interpreted in two ways: 1) empathy drives a symbolic relationship about how this feels and 2) sympathy drives a sign relationship about what this means.

Our epiphany in North Carolina at the FACE experiment was based upon a symbolic relationship between a tree, the sunlight, clouds and our interactions through plant physiology equipment. The experience gave a sense of lived connectedness. Stein calls it the 'phenomena of life [that includes] growth, development and aging, health and sickness, vigour and sluggishness' (2002, p. 68). She extends this idea further, 'we not only see such vigour and sluggishness in people and animals, but also in plants. Empathic fulfilment is also possible here' (Stein 2002, p. 69). With empathy, we not only understand, but we feel the other's health, well-being or emotional state. Empathic projection helps us to imagine ourselves as if the other is looking at us and judging our behavior. Lakoff and Johnson define it as an 'imaginative experience of the other' (1989, p. 566). This specific imagination is cued by the empathic relationship between what is perceived and the perceiver. Imagination works through metaphor to enable our understanding of the 'other' and the environment. This is key to an empathic approach to nonhuman living things.

The research method

The first artwork in the *Eden3* series, titled *Plein Air*, consists of a chamber holding a leaf. Air is pumped through the chamber and connects to high quality sensor technology embedded in a traditional painting easel with a computer processing equations that measure and sonify photosynthesis and transpiration. The measurements focus on leaf reaction and the reduction of carbon dioxide and the increase of humidity. What is 'heard' is the sonic representation of tree leaf photosynthesis and transpiration data (this is explained in more detail below). There is also a sensor-based indication of carbon dioxide and humidity exchange as the tree leaf stomata (thousands of small pores on a leaf) open and close in response to changes in local atmosphere and weather conditions.

With this artwork, we sought to respond to the theories and experiences discussed above and to explore cross species (tree–human) empathy and interrelationship. We did this through a focus on revealing reactions to shared environmental conditions. Trees are alive, yet perceived as non-reactive entities operating within a time scale at the edge of human perception. Humans affect our shared environment through anthropogenic production of carbon dioxide as a by-product of industry, transport and development, as well as by breathing. Yet people have little sensitivity to the local impact – the small-scale cause and effect we have on atmospheric conditions in places we frequent. A tree, on the other hand, can react to small changes in the amount of carbon dioxide in the air (in parts per million). *Plein Air* is an experimental approach to a relationship with another species that shares our everyday context. The research, in both its technical and artistic forms is focused on the reactions of trees and creating conditions where attention (guided by aesthetic experience) enables potential for empathic exchange over time.

Sensors and sounds

The intention was to embrace key elements of our experience in North Carolina and to develop a portable sculptural instrument that would support an attentive empathic experience with trees, and that would develop over time. We were working with the recognition of the speed of reaction as a tree leaf responds to changes to carbon dioxide and sunlight, and the move from a language-based visual-sign output to a sensory sound-symbol output. The challenge was how to retain the focus on a tree and its environment. The mediated experience with sensors and sound interrupts perception and the idea (the normative value) that trees are slow moving things out of sync with human experience of time and place. Following these ideas, the sculptural interface needed to work in real-time. The constantly changing environmental conditions and physiological response of the leaf needed to be closely synchronized with the sound interface, if the sensitivity of plants to atmospheric change were to be revealed.

Here we provide a brief overview of the project phases. The project was initiated when one of the authors, Goto, was offered a bursary for doctoral research at Robert Gordon University in Aberdeen, Scotland; six months later, Collins secured an interdisciplinary research grant from the University of Wolverhampton in the West Midlands in England. It has evolved through iterative development. In 2008, the project was conceptualized and initial sensors and plant materials purchased. The project functioned in terms of data collection and data sonification in two separate steps. Between 2009–2010, we spent three months running science/sound experiments at the Headlands Center for the Arts in California, where the sculptural form of the project was also developed (see Figs. 7.1, 7.2 and 7.3). Upon return to the UK, we took up residency in the University of Wolverhampton Crop Technology Unit where we worked with agricultural scientist Trevor Hocking and PhD researcher Mat Dalgleish to refine the scientific interface and establish a real-time sound system (see Fig. 7.4). During this residency, the second version of the sculptural form was built and we decided the sound interface would be based on standard synthesizer software. The project was then exhibited with funding from Peacock Visual Arts, Aberdeen Scotland (2010). In 2011, the project, along with a live-sound response, was presented at two university seminars and one group exhibition. From 2011–2012, we worked with Michael Baldock and Clare

Figure 7.1 The first iteration included the project as a table top experiment.

(All photos courtesy of Collins & Goto Studio).

Figure 7.2 A mock up development of the easel tested in California.
(All photos courtesy of Collins & Goto Studio).

Cullen, musician/composers with sound programming expertise, to experiment with specific compositions and sound systems. Chris Malcolm, an experienced programmer and alternative musician, came on-board in 2013 to help rethink the interface and rewrite the software. The sculptural form was redesigned including a new visual interface and additional high quality speakers. Major changes were made to the computer and electrical systems. The project became the centerpiece of an exhibition at the Tent Gallery at the Edinburgh College of Art (2013). Final

Figure 7.3 Detail of the leaf chamber.
(All photos courtesy of Collins & Goto Studio).

refinements were then made to the sound system and the form, and a new live leaf-video program was produced for an exhibition and performance in Cologne, Germany in 2015. As we will discuss below, a key issue for the project was how to sonify the data in order to create the empathic experience we were seeking.

The **first iteration** was a desktop full of plant physiological equipment used to collect data and a separate software sound system to play (and test) the data. We devised a series of creative experiments within these limitations during our residency at the Headlands Center for Arts in California. The **second iteration** was a real-time art/science field instrument, a sound easel run by batteries. To keep the project moving (with winter coming on) we needed plants to work with. The residency in the Crop Technology Unit provided input from a plant physiologist and access to light and climate-controlled chambers as well as a greenhouse. There we were able to refine and reengineer the components to create a functioning real-time sound system (based on the Windows OS sound synthesizer), and a final design for the easel. *Plein Air* had become a complex instrument; it was tuned and tested by an engineering consultant. A custom/portable hardwood easel was hand crafted. It was used in fieldwork and then for extended exhibition in Aberdeen, Scotland (see Figs. 7.5 and 7.6). A woodwind and a piano were chosen from the standard synthesizer library as the 'voice' for the sonifications. The piano voice was chosen for its data clarity; however it was criticized during the first exhibition

Figure 7.4 The moment we first hear real-time sound. Mat Dalgleish and Reiko Goto working in the Crop Technology Unit with the test chamber behind them.

(All photos courtesy of Collins & Goto Studio).

and subsequent seminars. The identity (the image of a piano as the source) of sound overwhelmed the tree. The woodwind was less jarring, since it had more conceptual congruity, but its data clarity was ambiguous. The visibility of the technology would also be examined. Changes would be made in the next iteration.

The **third iteration** was characterized as a robust musical instrument played by a tree. Sound artists, Michael Baldock and Clare Cullen sampled specific wood/

forest sounds for percussion and used a complex program to create an integrative (data-driven) rhythm track. Working from London, they developed a composition with five tracks. It was beautiful and captivating, but difficult to sense the changing physiology of the tree. We then met with Chris Malcolm, a Glasgow programmer known in the alternative music community. He listened carefully to what had been done. He decided to abandon the sampling idea and work with data-driven custom sound synthesizers and banks of resonators. Working closely with Goto and specific trees on weekends, he approached the work like a jazz musician trying out different programming riffs and refining them to hear the changes occurring with the tree over time. We agreed to develop a data/sound library the tree could play. We spent months in development, testing his programming once a week with different live trees. The effort began with a distinctive data-driven bass line that would reveal increases and decreases in photosynthesis and transpiration. Malcolm then developed a complex set of accompanying tones 'shaped' by subtle changes in specific sensor data. The final product was robust, providing the listener with a sense of the changing conditions through experience and imagination. It also required a new computer, and a new dedicated live leaf/data display. The new sound required better speakers; the weight and additional equipment drove another reconstruction of the *Plein Air* form. It had become less portable, but the sound was fantastic and, more importantly, the output reflected the changes to tree physiology while still retaining musical integrity.

Evaluating the empathic experiment

Work on *Eden3* was complicated by the fact that deciduous trees are in leaf only five out of twelve months per year in the UK. The system that was initially conceived as a simple group of sensors had also grown into a complicated array of electronics. Planned as a lightweight interactive sculpture with a laptop for outdoor use; it evolved into a robust musical instrument that required mains-power. Our practice moved from dynamic outdoor environment to a stable indoor environment in order to achieve clarity of function and purpose tied to the pursuit of empathic practice with trees.

The challenge was to develop artwork that would provide a reason for 'being with' trees. There were only two options for the development of this research to be with trees in nature, or for the trees to be with us. We worked with six potted trees (alder, ash, aspen, birch, hazel and oak) that were nurtured behind our flat in Glasgow. The practice of working with trees has changed from the experimental initial effort to managing the equipment and seeing patterns in the enquiry. Next was a long-term project of experimentation with the idea of 'being with' trees; trying to see the environment through the trees with the help of the real-time instrument. As *Plein Air* evolved over time, it has come to embody a certain self-contained gravitas as an object and an imaginative instrument to be played by a tree.

The painting easel was a metaphor unto itself, an important compositional element of the artwork. It provided a classic sculptural body and counter-balanced the scientific aspect. The easel was an early technology that embodies a history of visual attention to, and human expression of, the landscape and nature. This also connected the development of the artists' empathic ability to embrace and express the condition of the landscape 'other.'

We sought the input of colleagues and audience at every step of development. Some colleagues were curiously clear about what they would expect to hear from trees, many assumed that the sound would be soft and soothing. Others found themselves engaged by the sound-computer-tree relationship; they wanted to know the science and technology behind it. Still others got very caught up with the interrelational aspects of listening to the metaphorical sound, the breath of a tree. During the second iteration in Aberdeen, we worked outdoors with more than half of the fieldwork sites being urban. All were public spaces. The form and function of the sculptural interface engendered curiosity in these contexts, but little sustained interest. As we were both present, most people wanted to hear an explanation of what was going on. Maybe half would take the time to listen to the sound and consider the environmental context. These were indications of sympathetic engagement, in Stein's sense, where there was sensitivity to the experience, but it was processed through what was known, rather than resulting in a relational, imaginative projection which we were hoping to encourage. Installing the sculptural interface in the Aberdeen gallery (2010) within a greenhouse with trees was more effective. Without someone present to explain things, viewers would sit in a chair and relax as they took in the experience. Here there was more potential to engage the tree and the imagination at the same time. The more or less private space of the greenhouse supported engagement, but the mise-en-scène carefully developed for an intimate empathic relationship with a tree was compromised by the dominant sound-image conjured by the piano 'voice'.[12] The primary insight from attention to and dialogue with the audience at the exhibition was that sustained attention over time was a necessary element of the empathic exchange we were looking for.

During the exhibition in Edinburgh (2013), the project was shown on a raised platform with four trees; documentation of its development was presented on the walls (see Fig. 7.7). With the laptop replaced by a mini-computer, a graphic image was developed and presented on a flat screen. The audience could logically link the sensor display to the sound data; but this would again elicit more interest in the scientific meaning than the experiential feeling. Without the greenhouse enclosure, the audience had less impact on the atmospheric chemistry, and so that aspect of the experience was not as variable and somewhat less effective; although the new sound program was a significant improvement. The response to this exhibition was robust and cross-disciplinary (leading to additional opportunities). We came away with a clear sense of a final series of changes to the form, function and graphics.

The most recent presentation of the work (2015) was in a music venue, Neue Musik Koeln, in Germany (see Fig. 7.8). The audience included artists, musicians and composers. In preparation for the exhibition, Collins refined the structure of the easel while Goto redesigned the screen image. She also introduced a live-video image of a leaf where the color saturation was controlled by linking it to the

Figure 7.5 Detail of *Plein Air: The Ethical Aesthetic Impulse*, at Peacock Gallery, Aberdeen [2010].

(All photos courtesy of Collins & Goto Studio).

Figure 7.6 Detail of fieldwork with *Plein Air*, above the River Dee, Aberdeen.

(All photos courtesy of Collins & Goto Studio).

Figure 7.7 Eden3: Trees are the Language of Landscape, Tent Gallery at Edinburgh College of Art, Edinburgh [2013].

(All photos courtesy of Collins & Goto Studio).

photosynthesis rate. Malcolm developed the software, which resulted in an image that was incredibly beautiful and suggestive. We felt that the work, its sound, form and image were now producing an imaginative realm, where the viewer might grapple with the metaphor of the sound of a tree breathing. In Germany, people sat with the instrument and the tree; the impact was clearer. They also listened more carefully than a typical visual-arts audience. There was a lot of discussion about the tree and attention to the subtle complexity of the sound and the range of the instrument. In response, we are currently finishing a final proofing edit of the software and graphics. A touring version will include a vinyl recording and notational output. The installation will include photographs that support the material development of the project as well as the empathic response of the audience.

Exhibitions and seminars revealed issues demanding attention. Our ideas about the instrument and our role as artists seeking interrelationship with a tree was

Figure 7.8 Sound of a Tree, Curated by Georg Dietzler. In Visual Sounds: Bioakustische Musik, ON – Neue Musik, Cologne Germany [2015].

(All photos courtesy of Collins & Goto Studio).

rightly problematized. The technology would demand additional cost and complexity to achieve real-time experience. Then the form and function would be visually and experientially edited to insure the 'presence' of the tree itself. Sound and interface would be refined over and over again. Questions of ethics and appropriate representation would need to be addressed. Creative authorship, form and method would continue to move into the background, so the experience of being with a tree could come into the foreground.

Goto has spent her life as an artist working with non-verbal, animal and vegetal others. Creating a work that enabled an attentive space, where an empathic epiphany with a nonhuman other could arise was the fundamental intent of this work.

Conclusion

In this research, and following Stein, we have 'reached out' to grasp the vegetal other. As describe earlier, Goto identified the potential for empathic experience through attention to the 'vigour and sluggishness' in people, animals and plants. We have also found support for (what may be) a plant consciousness in the theories and science of Chamowitz and Trevawas. The *Plein Air* instrument has been developed and tested in a series of iterations. Results to date are cultural outcomes; although science is important to the work. Our collaboration with a plant physiologist establishes (on one level) the authenticity of our effort by supporting attention to the quality and method of gathering the data, as well as in the pursuit of programming that would let us hear the tree (through its physiological response). To 'hear the tree', we had to focus on that idea and control the foreground/background relationships. There is an implied truth and some confirmation of ethical intent in this intention and outcome.

The key question, as we suggest in the introduction, is this: Can artwork contribute to the evolution of a tree from a thing to a being of value? In a recent lecture, we argued that science (generally) contributes to our understanding of what a set of things are in form and function, while art focuses our attention on the individual qualities that sets people, places and things apart from common truths. The experience produced by *Plein Air*, mediated by sensors and software, lets us hear the (otherwise silent) metaphorical sound of one leaf/one tree breathing; does our sense of moral duty change as we listen? The tree is primarily (commonly) understood as property, as a utilitarian resource and as a non-sentient thing. The presence of trees in our daily lives and their bio-chemical agency can be construed as more public than private. Here the ecofeminist critique of a generative, reproductive body inappropriately subsumed by instrumental private interest and in need of emancipation has some traction (Plumwood 1993, p. 145; Merchant 1995, pp. 10–11; Salleh 1997, p. 29). Where moral duty is afforded to a tree, it is normally due to radical species loss,[13] or unusual cultural value.[14] Empathy is the leverage point that we have chosen to work with. We shaped the project over three iterations refining the form, the image and the sound to the point that it has become a simple instrument that sits between ourselves and one leaf, one tree. The experience designed to insinuate ethical duty.

To return to the question of participation that is the focus of this collection, we would argue that our project problematizes the discourse ethics of participatory research. This became particularly clear during our involvement in the *In Conversation with. . .* project, which is described in the chapters by Bastian and Heddon. We were invited to contribute to the final workshop that explored how a participatory ethics framework developed by Banks *et al.* (2013) might inform research with/on water. In reflecting on our discussions there, we would suggest that *Plein Air*, which skirts the edges of communicability and objectivity, calls the workshop's aims into question; the anthropocentrism of the ethics framework cannot simply be extended to human–nonhuman exchanges, whether this is with water, trees or other others. At best we can intend active listening and positive change. Hippocrates's phrase is useful in this work; '*primum non nocere*' first, do no harm. Our intentions cannot be matched, balanced or challenged by the trees we work with. The ethical charge of active hearing and making a difference are our sole responsibility. When we work with the nonhuman others, we are operating within a power differential that can only be checked by other human advocates. We know that everything we do has an impact upon perception, cognition, imagination and experience of the human-tree relationship. In the final form of *Plein Air*, we have repeatedly edited and reduced the material and technical presence to create an instrument that can be connected to a leaf for a tree to play. We have worked to produce an experience that focuses the audience on what Goto calls 'a sense of lived connectedness' (2012, p. 108).

Theodore Adorno has said, '[a]rt is not an arbitrary cultural complement to science but rather, stands in critical tension to it' (1997, p. 231). Working with *Plein Air*, the tree becomes the sensual other that we seek empathic interrelationship with to extend our experience and perception of the environment. The work follows ideas in science, and expands current theory and practice in our discipline. It raises questions about participative ethics in social science, planning and design. It offers a small challenge to the instrumental relationships we have to more-than-human life. Working in the realm of imagination and metaphor, this work requires deliberative agonistic response. The epiphany is to art, as discovery is to science.

Notes

1. The degree was awarded in 1916 at the University of Freiburgh in Breisgau, and the dissertation was published in 1917 at Halle, Germany.
2. Authors Claire Bishop (2004, 2012) and Nicholas Bourriaud (2002) provide a critical counterpoint to Kester, questioning the dialogic aesthetic by focusing on material output.
3. We commissioned Grant's work as editor and curator while working at the Studio for Creative Inquiry, at Carnegie Mellon University (see Collins and Goto 2005, pp. 10–15).
4. http://chrischafe.net/oxygen-flute/
5. http://www.acousticecology.org/dunn/solit.html
6. http://www.i-am-ai.net/work/the-dark-forest/
7. http://www.i-am-ai.net/work/a-conversation-between-trees/
8. http://www.livingsymphonies.com/
9. http://nmr.collinsandgoto.com
10. http://3r2n.collinsandgoto.com
11. For more information on the historic context, theory, method and initial work, see (Goto Collins 2012). Also a book chapter that looks at empathy across historic environmental art practices in (Goto Collins and Collins 2012).
12. This was confirmed in conversations with colleagues at a panel discussion during the exhibition, and then again at a follow on seminar and presentation of the system at the Institute for the Advanced Studies in the Humanities, at University of Edinburgh.
13. The common juniper of Scotland and Wales, for example, is considered a species of concern; protected under the Wildlife and Countryside Act 1981.
14. *Chinju no mori* or sacred groves around Shinto shrines are protected as the dwelling place of the *kami* spirits, although it is the *kami*, which are worshipped, the trees are the cultural ecology which supports their presence.

References

Adorno, T. W., 1997. *Aesthetic theory*. Minneapolis, MN: University of Minnesota Press.

Askins, K., and Pain, R., 2011. Contact zones: participation, materiality, and the messiness of interaction. *Environment and Planning D: Society and Space*, 29 (5), 803–821.

Banks, S., *et al.*, 2013. Everyday ethics in community-based participatory research. *Contemporary Social Science*, 8 (3), 263–277.

Bishop, C., 2004. Antagonism and relational aesthetics. *October*, 110 (1), 51–79.

Bishop, C., 2012. *Artificial hells: participatory art and the politics of spectatorship*. Brooklyn, NY: Verso Books.

Bourriaud, N., 2002. *Relational aesthetics*. France: Les presses du réel.

Burgess, M., 2011. On being moved: sympathy, mobility, and narrative form. *Poetics Today*, 32 (2), 289–321.

Chamovitz, D., 2012. *What a plant knows: a field guide to the senses*. Melbourne and London: Scribe Publications.

Collins, T., and Goto, R., 2005. Initiating the groundworks exhibition. *In*: G. Kester, ed. *Groundworks: environmental collaboration in contemporary art*. Pittsburgh, PA: Regina Gouger Miller Gallery, Carnegie Mellon University, 10–15.

Cooke, B., 2001. The social psychological limits of participation. *In*: B. Cooke and U. Kothari, eds. *Participation: the new tyranny?* London and New York: Zed Books, 102–121.

Cooke, B., and Kothari, U., eds., 2001. *Participation: the new tyranny?* London, New York: Zed Books.

Demos, T. J., 2010. The politics of sustainability: contemporary art and ecology. *In*: A.S.S. Witzke and S. Hede, eds. *Rethink: contemporary art and climate change*. Aarhus, Denmark: Alexandra Institute, 53–58.

Demos, T. J., 2012. Gardens beyond Eden: bio-aesthetics, eco-futurism, and dystopia at DOCUMENTA (13). *The Brooklyn Rail* [online], 4 October. Available from: http://www.brooklynrail.org/2012/10/art/gardens-beyond-eden-bio-aesthetics-eco-futurism-and-dystopia-at-documenta-13 [Accessed 15 October 2015].

Demos, T. J., 2013. *The migrant Image: the art and politics of documentary during global crisis*. Durham, NC: Duke University Press.

Demos, T. J., 2015. Rights of nature: art and ecology in the Americas. Nottingham: Nottingham Contemporary.

Deutsche, R., 1996. *Evictions: art and spatial politics*. Cambridge, MA: MIT Press.

Eubanks, V., 2009. Double-bound: putting the power back into participatory research. *Frontiers: A Journal of Women Studies*, 30 (1), 107–137.

Firn, R., 2004. Plant intelligence: an alternative point of view. *Annals of Botany*, 93 (4), 345–351.

Friedmann, J., 1987. *Planning in the public domain: from knowledge to action*. Princeton, NJ: Princeton University Press.

Frodeman, R., Briggle, A., and Holbrook, J.B., 2012. Philosophy in the Age of Neoliberalism. *Social Epistemology*, 26 (3–4), 311–330.

Gablik, S., 1984. *Has modernism failed?* New York: Thames and Hudson.

Gablik, S., 1992. *The reenchantment of art*. New York: Thames and Hudson.

Goto Collins, R., 2012. *Ecology and environmental art in public place, talking tree: won't you take a minute and listen to the plight of nature?* Thesis (PhD). Robert Gordon University, Aberdeen, Scotland.

Goto Collins, R., and Collins, T., 2012. LIVING things – the ethical, aesthetic impulse. *In*: E. Brady and P. Pheminster, eds. *Transformative values: human-environment relations in theory and practice*. London: Springer-Verlag, 121–133.

Harrison, H. M., and Harrison, N., 1993. *The serpentine lattice*. Portland, Oregon: Douglas M. Cooley Memorial Gallery.

Kastner, J., and Wallis, B., eds., 1998. *Land and environmental art*. London: Phaidon Press.

Kester, G., 2004. *Conversation pieces, community and communication in modern art*. Berkeley, Los Angeles, CA and London: University of California Press.

Kester, G., ed., 2005. *Groundworks: environmental collaboration in contemporary art*. Pittsburgh, PA: Regina Gouger Miller Gallery, Carnegie Mellon University.

Kester, G., 2011. *The one and the many: contemporary collaborative art in a global context*. Durham, NC: Duke University Press.

Koh, J., 2015. *Art-led participative processes: dialogue and subjectivity within performances in the everyday*. Thesis (PhD). Academy of Fine Arts, University of the Arts Helsinki, Finland.

Kwon, M., 2002. *One place after another: site-specific art and locational identity.* Cambridge, MA: MIT Press.

Lakoff, G., and Johnson, M., 1989. *Metaphors we live by*. Chicago and London: University of Chicago Press.

Lippard, L., 1983. *Overlay*. New York: Pantheon Books.

Lippard, L., 1997. *The lure of the local*. New York: New Press.

Lippard, L., 2007. *Weather report exhibition catalogue*. Boulder, CO: Museum of Contemporary Art.

Lippard, L., 2014. *Undermining: a wild ride through land use, politics, and art in the changing west*. New York: New Press.

Matilsky, B., 1992. *Fragile ecologies: contemporary artists' interpretations and solutions*. New York: Rizzoli International Publications.

Plumwood, V., 1993. *Feminism and the mastery of nature*. London: Routledge.

Salleh, A., 1997. *Ecofeminism as politics: nature, Marx and the postmodern*. London: Zed Books.

Sonfist, A., 2007. Public monuments [online]. *In*: *Proceedings from Artful Ecologies, Art, Nature & Environment Conference 12–15 July 2006, Falmouth 2006*. Falmouth: Research Art, Nature and Environment (RANE), University College Falmouth, 9–11. Available from: http://rane.falmouth.ac.uk/pdfs/artful_ecologies_papers.pdf [Accessed 15 November 2015].

Stein, E., 2002. *On the problem of empathy*. Washington, DC: ICS Publications.

Stueber, K., 2014. Empathy [online]. *In*: E.N. Zalta, ed. *The Stanford encyclopedia of philosophy*. Available from: http://plato.stanford.edu/archives/win2014/entries/empathy [Accessed 15 November 2015].

Trevawas, A., 2003. Aspects of plant intelligence. *Annals of Botany*, 92 (1), 1–20.

Trevawas, A., 2005. Green plants as intelligent organisms. *TRENDS in Plant Science*, 10 (9), 413–419.

Willisch, S., and Heimberg, B., eds., 2007. *Joseph Beuys: the end of the 20th Century*. München: Dorner Institute.

8 Empowerment as skill

The role of affect in building new subjectivities

Anna Krzywoszynska

Introduction

When I arrived at my fieldwork location at Colli Verdi in winter 2008, I undertook the task of vine apprenticeship; I was to learn how to work with and to care for vines.[1] This effort was part of my doctoral research: a more-than-human ethnography of organic wine making in Northern Italy (Krzywoszynska 2012). The immense awkwardness and alterity of facing a vine for the first time in this context and asking: 'who are you, and so who are we? Here we are, and so what are we to become?' (Haraway 2008, p. 221) was one of the pivotal moments of my research, and led me to engage with feminist and materialist writers as I sought to understand both this event, and the developing relationship between the vines and myself. In this chapter, I suggest that the process of becoming skilled in vine work can be thought as an emergence of a new self, understood after Haraway (2008) as an open network of meaningful relations. Haraway's perspective on the 'self' moves away from the idea of an inherent and fixed identity, and towards relational and mutable subjectivities that emerge from, and are constituted by, relations with human and nonhuman others. In this reading, skill can be seen as a reworking of a self through the development of new, meaningful relations with animate and inanimate nonhuman 'others'.

This understanding of 'self' as emergent from relations, practices and discourses has been taking root in debates and practice within participatory research (PR), particularly in relation to empowerment, understood as the capacity of individuals and groups to act according to their will. There has been a shift towards more-than-human relationality in PR debates, with explorations of the role of space, embodiment, materiality and affect. For example, social scientists Caitlin Cahill (2007), Amanda Cahill (2008) and Margaret Morales and Leila Harris (2014) have called for an understanding of empowerment as an effect of situated relations as much as an individual capacity to act on these relations. In a similar vein, participatory action scholars Mike Kesby et al. (2007) call for a reconceptualisation of empowerment as a spatial practice, and as an effect of the deployment of certain resources. In this chapter, I contribute to this debate by exploring the fruitful resonances between empowerment and enskillment (Ingold 2011), and I propose that empowerment can be usefully seen as a skill in and of itself.[2]

Figure 8.1 Winter vines as aesthetic objects.
Photo by the author, January 2009.

Both empowerment and enskillment require cultivating new forms of subjectivity through active work on the relations which constitute a 'self'. I suggest the work of building new relations requires cultivating unfamiliar affects. I use an auto-ethnographic account of acquiring vine work skills to explore the affective states cultivated in my vine apprenticeship, and to highlight the potential resonances between the roles of affects in developing vine working skill and in developing empowerment. The practice of auto-ethnography requires a critical awareness of the effects of research on the self, and is thus well suited to an exploration of the work of building new relations through practicing unfamiliar embodiments and affects, and to reflecting on the socio-material conditions which make these processes possible (O'Connor 2007). In the discussion, I explore some ways in which the affective states that enable a reconfiguring of the relational self may be used in PR practice for cultivating empowerment.

Relational subjectivities in developing skills and empowerment

The concept of skill as a transformative dialogue between the doer and the world arguably goes back as far as Aristotle's *Nicomachean Ethics* (ca. 350 BCE), and has seen a recent revival in social sciences through an interest in craft championed by the sociologist Richard Sennett (2009), and in relationality of action and perception by the anthropologist Tim Ingold (2000, 2011). These works have stressed the centrality of sensing and sensation for the interaction between mind-body and its environment in the process of learning (see also Crossley 2007, Gieser 2008, Lea 2009). Being skilled means being aware of the ways in which the world of action unfolds and understanding one's capacity to participate in this unfolding. My understanding of the character of this adjustment between the doer and the world also draws on relational materialism (Anderson and Harrison 2010), which sees bodies and worlds co-creating one another. Human bodies are transformed from a self-contained vessel *of*, and tool *for*, the intellect into 'a dynamic trajectory by which we learn to register and become sensitive to what the world is made of' (Latour 2004, p. 206). Similarly, the material world is no longer a Euclidean plane populated with separate, passive 'things', but instead a dynamic, relational unfolding of materials and forms (Ingold 2000). The notion of affect is central to this perspective. After geographer Jamie Lorimer (2007), I understand affects as a collection of shared and interconnecting forces operating between bodies. The relationship between affect and affordance is also important to stress. Affordance is a relational quality, arising from the meeting of 'the inherent, ecological characteristics of a nonhuman in relation to the phenomenological apparatus of the body (human or nonhuman) that encounters and perceives them' (Lorimer 2007, p. 914). Different bodies present different affordances, and so the recognition and responsiveness which structure affective states do not depend exclusively on the efforts of a (perceiving, interacting) actor, but are influenced both by the perceptual apparatus of the actor, and by what the other who is interacted with presents.

Consistently with this literature, in this chapter I suggest that developing skills can be understood as a process of developing new subjectivities, drawing on Haraway's (2008) understandings of subjects as relational, mutable, and changing. Haraway further stresses the self is always already a 'multispecies crowd' (2008, p. 165), with other species being both the condition of existence, and the co-creators of meaning. This perspective rejects the notion an original 'I' to which experience is 'added', but rather sees the self as arising from the entanglement in relations with the human, nonhuman and elemental others (see also Latour 2004). The self is then understood as a product of mundane and everyday relationality and, being open to this dance of relating (Haraway 2008, p. 25), is understood as how we are in the world – not a bolt-on, not an extra. This multi-species relational perspective rings particularly true in the case of grape-producing vines and wine-drinking humans, as their co-dependence and co-entanglement runs deep. The phylloxera outbreak in late 17th century forever bound European grapevines (*vitis vinifera*) to humans, as to survive they began to be grafted onto the pest-resistant American vine (*vitis lambrusca*) rootstocks. Since then, a complex system of control and certification has developed which governs the circulation of vines as bio-technologies, grown at certified nurseries out of clones developed and owned under intellectual property rights (see Krzywoszynska 2012, chapter 4).

Such relational understandings of the self as a tangle of relations have been gaining prominence in debates on empowerment in participatory research. 'Sovereign' perspectives on empowerment which saw people as 'possessing' empowerment and therefore 'holding' a capacity to act in particular ways have been critiqued by a number of authors (Kesby 2005, Cahill 2007, Cahill 2008, Morales and Harris 2014, Wijnendaele 2014). In turn, the relational, spatial and temporal natures of subjectivity have been highlighted, stressing the self as a constant work in progress (Kesby 2005). They have resulted in empowerment being re-cast as a change in the composition of one's self, enabling one to re-engage with the world in a new, and hopefully more positive or productive way. These perspectives highlight that empowerment is not something one acquires, but is something one becomes; it cannot be 'given', but has to be developed by each individual and group as a new way of being in the world. As a result, some participatory research processes have experimented with creating spaces and times for the exploration of one's multiple-situated positionalities, and for the development of new subjectivities – an 'iterative long-term shifting process of learning, making sense of one's subjectivity, and reworking it through collective dialogue, ongoing reflection, and analysis' (Cahill 2007, p. 276).

In the following sections, I draw on these relational perspectives on the self and on experience to reflect on the role affective states played in my process of acquiring vine work skills. I explore the three key affective states which marked the change in myself from an 'aesthetic observer' to a 'pruner' of vines: enchantment, becoming and focus. Enchantment allowed me to see the possibility of relating to vines otherwise; becoming (working) vine allowed me to bring temporally and spatially distant events and bodies (of significant humans and nonhumans) to bear on my learning experience; focus indicated the coupling of action and perception that is characteristic of skill. I explore how PR methodologies may draw on the role of affects in skill acquisition to aid participants' reconfiguring of their subjectivities towards empowerment.

Enchantment

To create meaningful relations one needs some basis for connection. While this connection is easier to establish with some nonhumans more than others, enchantment has the power to create connections across even radical difference (Bennett 2001, Lorimer 2007). My previous engagements with plants as living beings in the landscape have been predominantly aesthetic, and during my winter fieldwork at the vineyard I continued to focus on vines as aesthetic objects, as illustrated by photographs I took at that time, such as Figure 8.1. Initially, I was unavailable to the affects of vines as objects of work. Forging new relations is not easy. We often try to reproduce familiar patterns, and during my first engagements with vines in the winter season I struggled to move beyond the position of an external observer as, following and talking with other workers, I bore witness to pruning work. I struggled to understand how pruning choices were made as the workers cut deep into the tissues of the plants. My quest for what I then saw as a hidden meaning behind the actions deployed the 'engineer' model of understanding skill, critiqued eloquently by Ingold (2000), which presumes a set of rules pre-existing material engagement, and directly imposed onto the world. As I have explored in more detail elsewhere (Krzywoszynska 2015), my early meetings with vines were thus dominated by the idea of a repository of meaning separate from practice, which the workers 'translated' into the body of the vine. This idea was reinforced by diagrams such as Figure 8.2, which I took to be proof of a 'universal truth' of vine pruning. For me, winter was a time of frustration, where I started to understand in abstraction why certain things were done, but struggled to connect the intellectual understanding with a practical one.

To become skilled at vine work, I needed a way to open myself to the world of action, and to the vines and vineyards in which I laboured, and to stop struggling to connect the intellectual understanding with a practical one. I had to let go of the idea of pre-existent meaning; instead, I had to discover the meaning in situ, forge my own way of going forward. When the spring came, I was drawn to the changing bodies of the vines, exploding with sensual greenery; it seemed I could practically see the branches grow. I wrote in my diary that:

> *Shoots are lovely to touch . . . Sensually, it is a completely different experience to winter pruning. Before, I was struggling. Now, on the contrary, I have to pay special attention and be extra delicate to make sure I don't do damage (Peter told me I ought to 'caress the vine' at this stage). (. . .) There is more to see now, the vine seems more alive, and it is easier for me to start to think about the force it will need to create grapes, about how many grapes it can support, which branches it can develop, see it as a totality, a living thing. The shoots are a beautiful light green, and they are extremely vulnerable and brittle, they pop off the branch at the most delicate touch. You hardly need tools, we work with our hands. Soon they get covered in fragrant vine juice, it smells lovely, a fresh, green smell.*
>
> *(field diary 05/05/2009)*

The affordances of the spring vines indicated to me a different way of relating to the vines than I felt was possible when faced with the inert 'sticks and knots' of their winter form. The sensual enjoyment opened up something in me; the vines touched me in a new way and demanded my attention. Enchantment is memorably discussed by Bennett (2001) as a state of wonder, a temporarily immobilising encounter characterised by heightened sensibility and exhilaration. Enchantment brings into awareness something which had not been noted before, and so opens up the possibility of relating. The plants' enchanting qualities called upon me to respond, and so enrolled me into new relations (Hitchings 2003). Sensual enjoyment offered a way in, which was further shaped by practice with my vine working colleagues.

Becoming (working) vine

My attention was directed toward the spring vines not only for sensual enjoyment, but towards the goal of 'green pruning' (also known as canopy management), in which superfluous branches, leaves and grape buds are pruned to further aid the development of high quality grapes. This learning was aided by the materiality of the spring vines, which in some respects are more forgiving of

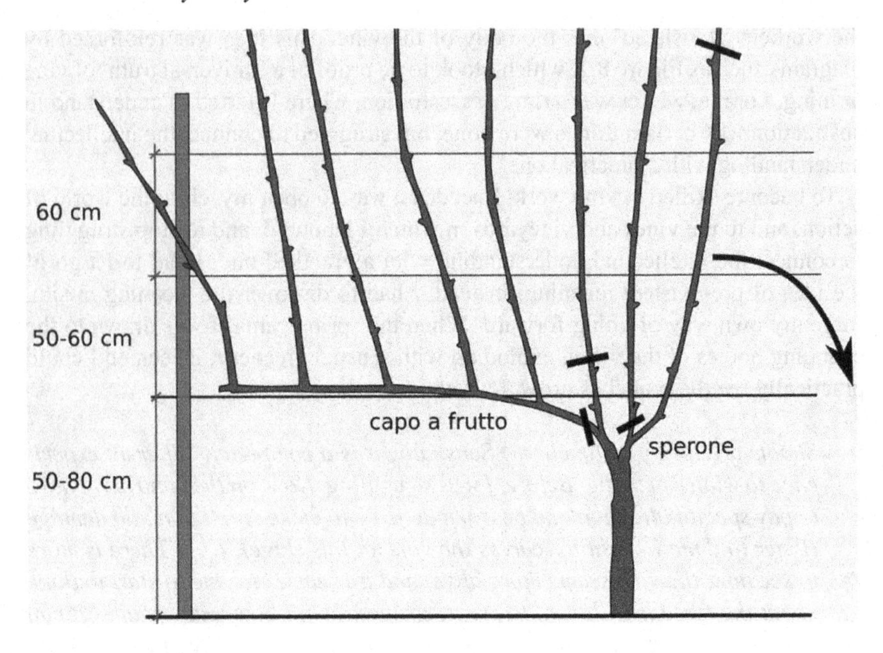

Figure 8.2 Schematic of a Guyot trained vine with pruning cuts indicated (Giancarlo Dessi 2008, http://commons.wikimedia.org/wiki/File:Guyot_1.svg).

mistakes than winter ones, as they continue to vigorously produce shoots. Avoiding mistakes, however, required a reconfiguring of the way I habitually attended to plants. I had to shift my perception of vines from mute and stable objects in front of me, to seeing them as temporally unfolding and spatially connected entities: as living others, always already entangled in relations with human and nonhuman others.

Discovering significant relationalities of the vines was challenging as they presented an ethology very different to my own, inhabiting time and space differently to the way I do (Jones and Cloke 2002). Rooted in place, vines exist through their mutually impactful relations with others, elemental, human and nonhuman, enrolling them to further their growth and reproduction (Atchison and Head 2013); they are also highly adaptable and sensitive to their environments. Vines are indeed so entangled in their environments as to challenge simple distinctions between individuals and collectives, landscapes and their constituents. Their seasonal temporality can be both off-putting and overbearing. Vines appear mute, seemingly dead in winter, while in spring and summer, their capacity for unassisted and vigorous growth can challenge attempts at control and containment (Hitchings 2003, Barker 2008, Ginn 2008). The longevity of vines further confused my attempts at relating. Vines can be affected by embodied memories of events long past, or events so momentary as to be inaccessible to an untrained, uninformed workers' perception.

Other, more experienced workers helped me to tune by body and my feeling to what was significant to the vine as a relational, but also productive, organism – to what was significant, not only to secure the vine's health, but also to secure a desirable crop. The affect that was being cultivated in me was therefore what, drawing on the philosophy of Deleuze and Guattari (1988), authors in geography have described as the work of 'becoming animal' (Lorimer 2008, Bear and Eden 2011), or in my case 'becoming (working) plant': a different way of resonating and reacting to the world around me. I was seeking to become not just any plant, however, and Bear and Eden (2011) are right in drawing attention to the important specificity of the nonhumans we seek to get close to. The work of situating myself in relation to the being of the vine had as much to do with understanding how my body and my actions interacted with the bodies and activities of vines (growing, reproduction, decay), as with the bodies and activities of other workers (pruning, spraying, harvesting). This meant developing new ways of feeling and being with the plants, in the group and in the landscape, cultivating in my body the capacity to react to signals and configurations which had hitherto passed unnoticed.

Reconfiguring my habitual ways of looking and inhabiting place were central to this work. On one occasion, Damian, the team leader, brought me back to a row that I had just worked to correct my performance. It was early summer, and mildew was starting to develop. In this organic vineyard, sulphur sprays were used to protect the plants (instead of more conventional fungicides). Damian asked me to kneel beside him, and to look along his extended arm, into the vine and across, up the row. Leaves were blocking my view in all directions; this meant there was no air circulating, and the sprays couldn't penetrate either. 'You have to really get in there', Damian told me, as he cleared the surplus, creating a new, airy feel. 'That's better. Now you try'.

'You have to be able to see through the vine' was just one of the myriad rules of thumb I was taught during my vine apprenticeship, and there were always more to learn. Rules of thumb were not instructions: This is what must be done. They were didactic tools: Attend to this, and use your own judgement (Krzywoszynska 2016). They encouraged me to embody the space differently, as part of a network of past, present and future relations in which my activities played only one part. Are there too many leaves here, and if so, which ones should I remove? Leaving young and pruning old to prevent mildew, leaving midday leaf 'umbrellas' to shade the developing grape bunches, and pruning the shoots growing too close together to make harvesting easier for the future workers required me to develop a sensitivity to the vines and to other workers. Rules directing my attention were 'like the map of an unfamiliar territory (. . .) the map can be a help in beginning to know the country, but the aim is to learn the country, not the map' (Ingold 2000, p. 415). The aim, for the human workers there, was to make wine. This involved acquiring an embodied experience of the unfolding of vines in the processes of vine work, which required ongoing, and long-term, cultivation of attention to the temporally unfolding materiality of vines and vine work.

In addition to rules of thumb, the education of attention in vineyards frequently incorporated the telling of (hi)stories. Like rules of thumb, they helped communicate a change in practice amongst experienced workers, and to direct the attention of the less experienced ones so that the knowledge expressed could be rediscovered by the novices as they engaged with the vines (see Ingold 2011, p. 159). They made it possible to situate green vine pruning in the longer lifecycle of a vine and of a vineyard: the moment of encounter in the dance of relating. In the context of this historical knowledge, in each vineyard, pruning was re-evaluated, as seen in Damian's description of *Vigna Nuova*, where vines were being trained to produce high quality grapes:

> *Damian: Last year, for example, there weren't that many grapes, but all the same we went through the vineyard and plucked bunches off to make the vineyard used to making fewer. And, in fact, it is already making fewer. (. . .) This vineyard always has to produce little, and of high quality. It is different with Barbera in Vigna di Carla – the terrain is more flat, and there is less clay, so we know the grapes will always be less good, so we leave more on, for the demijohns, or less expensive bottles.*
>
> *(field diary 07/05/2009)*

Such (hi)stories, like rules of thumb, allowed the more experienced workers to communicate a shared history of embodied experience of these vines. They brought together what was in front of us now, what came before, and what may come after. Vines grow slowly, which can result in epistemic distance for the working human (Carolan 2006): not all that is relevant can be perceived at one time. Similarly, many relations are hidden to the casual observer: Although the character of the soil and the (hi)story of the weather and the vines were not perceptible for me, through the telling of the vineyard's stories these materialities were made visible and their significance more clear. In addition to the thinking-backwards and thinking-forwards, to where the vine had been and where it was going, we were also encouraged to think 'sideways', considering the relations of which the vine was part, such as those with soils or the weather. These relations were always materialising in the bodies of vines. By being attentive to their material presences and appreciating their power, I could become more sensitive to how my relationship with the vines (through pruning) fitted in the wider mesh of relations. Enrolling the vines into this mesh of relations which made my 'self' brought with them soils, water tables and other workers. I was not making relations with 'any' vines, with abstract entities, but with the very real plants rooted in the space and history of a particular place.

Focus and its rhythms

Enchantment provided a way in to the network of relations around a working vine, and attempts at becoming-working-vine made the relations available through the training of attention. Exercising the skill, in turn, required focus, a state of being 'in the zone' described in studies of skill as maximum grip (Merleau-Ponty and Smith 1996), flow (Csikszentmihalyi and Csikszentmihalyi 1992) or being in harmony in the world of action (O'Connor 2007). In the context of leisure skills such as parkour (Saville 2008) or rock climbing (Thrift 2008), this affect is often described as a seamless coupling of action and perception where the 'I' disappears and there is no sense of effort. In the context of labour skill, however, this state, I found, is not effortless. In contrast, staying focused is tiring and requires work, as I recall in my research diary:

> *When working the rows, we become fully submerged in the plants. Hardly anyone talks, and we all concentrate hard on not making mistakes. Every few rows, every hour or so, we take a break, sit on the grass for a quarter of an hour and smoke and drink water. Some stretch out on the ground to give the back and the arms a rest, after all we are spending the day bent in two. (. . .) [the wine-maker] walks past and jokingly tells us off for loafing about. Damian waves him on, but gets quite defensive after a while. It is hard work, he says, lying in the shade of the leaves. The winery workers don't get it, but it requires so much concentration, it is intellectual as well as physical. If you let tiredness take over, and you start going on auto-pilot, you can do a lot of damage.*
>
> *(field diary 09/05/2009)*

In my developing of vine working skill, affordances and affects did not come effortlessly together. Playing my part of the relational network well required work. Mistakes happened, not only to me, but to other workers, as tiredness made us numb to the world around us, and caused us to switch off. These experiences problematized for me the notion of an effortless connection between doer and deed in the exercise of skill. While attentiveness to the rhythm of the world of action has been noted as an element of skilled practice (Sennett 2009, Bear and Eden 2011, Ingold 2011), the rhythm of the focus itself is less commented on. Every skilled practice, however, carries an inherent risk of practice breakdown (Harrison 2009). I saw that the work of even the most experienced practitioner has a wavelike temporal quality in which peaks of attunement change into lows of disconnection. This rhythm of skill was written into the very unfolding of vine work practice, the intensity of which we could not sustain for long periods of time, or in the difficult conditions of midday heat. Relating well meant also knowing when to let go and recover, appreciating the inherent effort.

Working on affects for cultivating empowerment

Reconfiguring myself to develop meaningful relations with working vines was aided by affective states which made the connection possible (enchantment), which stretched my awareness (becoming (working) vine), and which indicated things were going well (focus). These affects, I suggest, can be usefully reflected on and utilised in PR methodologies which aim to support the development of empowerment understood, like skill, as a reconfiguring of one's relational engagement with the world.

Becoming otherwise through the cultivation of attention has emerged as an important theme in recent PR scholarship on cultivating empowerment. Scholars working from the perspective of relational subjectivity propose that PR projects should offer spaces for critical analyses of one's situatedness within broader social, spatial and political processes (Cahill 2007). This includes exploration of the ways in which one has embodied and internalised potentially oppressive power relations (Kesby et al. 2007). Drawing conscious attention to the norms governing one's life and subjecting them to scrutiny can make one aware of how one's everyday choices reproduce power-laden relations, and it can be an important step toward changing the subjectivities prompted by the norm (Morales and Harris 2014). This explicit exploration of one's lived entanglements – of one's subjectivity – is facilitated by PR practitioners, who seek to support the generation of new narratives for social relationships (Morales and Harris 2014), and to construct the space of participation as a field within which 'opportunities open up for people, first, to disentangle the complex web of everyday life (. . .) second, to deconstruct norms and conventions; third, to reflect on the performativity of everyday life; and finally, to rehearse performances for alternative realities' (Kesby 2005, p. 2055). For example Catilin Cahill, in her work with young working-class women of colour, used PR processes to encourage her participants to reflect critically upon their everyday lives as part of a collective process which emphasised dialogue, and in so doing helped them rework and redefine their subjectivities (2007, p. 273).

While cognition has been stressed as a source of action in developing new relations of empowerment, the role of materiality, embodiment, and affect in building empowered subjectivities has also been noted, as discussed above, and can be built on further. It is in this context that the affects of enchantment, focus and becoming can play a role in creating and sustaining new relations, leading to engaging with the world differently. Authors have stressed that cultivating new ways of being in the world involves developing a whole new affective and bodily praxis (Wijnendaele 2014); playing on these new embodiments through affects may offer powerful ways of exploring how different relations feel. Some PR practitioners have sought to create opportunities for such 'rescripting of interactions' (Boal 2000, Morales and Harris 2014) through theatre (Boal 2000), role-playing (Kesby 2005), participatory art (Askins and Pain 2011), participatory video (Hume-Cook et al. 2007) and photovoice (Krieg and Roberts 2007). Through such work, participants have a chance 'to "do" and "feel" things differently; by offering them a direct embodied experience/experiment of how things can be/feel different' (Wijendaele 2014, p. 277). Using these methods and others, moments of enchantment, in which the habitual way of being is suddenly and exhilaratingly shifted, of becoming, where new ways of resonating with the world are experimented with, and of focus, where perception and action become coupled, can surely be created. Thus these established methods can be pushed further to support the cultivation of new subjectivities not only through rational discourse but also by developing new affective registers.

Conclusions

Pregnant with her second child, a friend of mine commented once: 'when you're expecting, the whole world is pregnant!' Suddenly, she could see pregnant women everywhere she looked. This sensation is familiar to all of us (although not necessarily through the experience of pregnancy!); our embodied histories change our engagement with the world, stretching us out, making more – or less – of the world available to us. If you've just started gardening, the whole world is made up of gardens; if you've just read Marx, class struggle is everywhere. Instead of understanding this as a change in perspective, as if our 'self' were putting on a new pair of glasses, we can follow Haraway (2008) and re-ground selfhood in material practice. If we reject the idea of an independent, pre-existent mind, and look at the mind-body as a dynamic trajectory (Latour 2004) and made through the relations of which it forms part, acquiring skill becomes not so much the adding of a capacity as becoming someone new (perhaps we know this intuitively when we say 'I am an academic' or 'I am a driver'). The skill does not simply 'appear'; by experimenting with and being worked by affordances, we develop new capacities for relating which may result in skill.

To date, in PR debates, one way of understanding the acquisition of skill (of public speaking, of IT literacy etc.) has been in a supportive role to the work of empowerment. However, I suggest we can also usefully understand empowerment itself as a skill: that is, as a goal-oriented praxis. This gives empowerment a translatability which allows it to go beyond any immediate context of a project or issue, and to be enacted in various contexts. While the stress is different, there are striking similarities with regard to the basic dynamics of skill and empowerment, both in terms of their acquisition, and in terms of how they are lived and enacted. Both skill and empowerment are enacted by individuals, but emerge from a field of social and material relations. Both require a reconfiguration of relations between one's mind-body and those of others, human and nonhuman. This reconfiguration requires both training attention towards one's existing relations, and developing capacities to act in a new ways to cultivate new relationalities. To achieve this, both empowerment and enskillment benefit from 'other spaces' governed by discourses and practices quite unlike those that order everyday space and agency (Kesby 2005). In these spaces, new affective registers can be cultivated which enable and sustain the creation of particular relations.

In the case of vine working skill, the exercise of skill requires a positioning of individual mind-body in relation to others who constitute the realm of action – workers, plants, soils, chemicals, vineyards etc. In the case of empowerment, the object is similar. In both cases, the work involved with 'rescripting of interactions' (Morales and Harris 2014) requires a training of attention. Such attention cannot be only conscious and critical, separate somehow from the material environment of action, but must also be embodied and felt. With interaction between human and nonhuman others in a supportive setting, we experiment with new ways of feeling and being, with new affective states. Affects are thus a gateway to new relationalities. Enchantment opens up new affordances in the world around us – new opportunities for meaningful engagement. For me, the training of new affects resulted in vines morphing from aesthetic 'knots and sticks' into sites of conscious engagement and work. My hands knew what to do, as indeed did my entire body; while I can't claim to have become fully competent, I knew I was moving in the right direction. Similarly, for Askins and Pain's (2011) young participants, experience of playing together within the space of the PR project resulted in new friendships across hitherto insurmountable lines of difference.

PR practitioners already use a range of techniques to aid in the reconfiguration of subjectivities and exploration of how life can be felt differently. The importance of affect in developing new relationalities with humans and nonhumans reported in this chapter suggests that a further focus on affect in reconfiguring selves for empowerment is a valuable direction of travel. By approaching empowerment as a form of skill, PR practitioners can further reflect on what 'empowerment apprenticeship' may entail, and what conditions may be necessary or desirable for exercising empowerment beyond the spaces of PR encounters. As Keby et al. note, PAR epistemologies and methodologies allow participants and researchers to achieve new forms of agency, however these need to be 'constantly redeployed and normalised if empowered performances are to become sustainable' (2007, p. 24). Understanding empowerment as skill draws attention to its relational character; it does not come into being ex nihilo, but its development is aided by particular material and social conditions, as well as repetition. Empowerment is also not carried by an individual as a stable construct, but emerges from a field of relations, and constantly is responsive to the social and material context in which it is utilised. Furthermore, studies of skill can be helpful for PR in thinking beyond the human-to-human dimension of becoming empowered, and inviting an interrogation of the material and more-than-human relations which come into play. Crucially, understanding empowerment as a skill stresses that it is not something that can be given or received; it has to be lived and felt. Like skill, power exists only in its use. To paraphrase John Allen, power is no more to be found 'in' the apparatus of rule than skill of playing is to be found 'in' the wood of a violin (Allen 2003, p. 5). By helping one another cultivate new affective relations within the world, we can explore the skill of playing the violin of power to our own tune.

Notes

1 On the question of care in working with plants, please see Krzywoszynska (2016). All places and persons in this article have been given pseudonyms.
2 Drawing primarily on Ingold (2000, 2011) I choose to speak of 'skill' rather than 'craft'; however I acknowledge these terms are increasingly used synonymously.

References

Allen, J., 2003. *Lost geographies of power*. Oxford: Blackwell Publishing.

Anderson, B., and Harrison, P., 2010. *Taking-place: non-representational theories and geography*. Farnham: Ashgate.

Askins, K., and Pain, R., 2011. Contact zones: participation, materiality, and the messiness of interaction. *Environment and Planning D: Society and Space*, 29 (5), 803–821.

Atchison, J., and Head, L., 2013. Eradicating bodies in invasive plant management. *Environment and Planning D: Society and Space*, 31 (6), 951–968.

Barker, K., 2008. Flexible boundaries in biosecurity: accommodating gorse in Aotearoa New Zealand. *Environment and Planning A*, 40 (7), 1598–1614.

Bear, C., and Eden, S., 2011. Thinking like a fish? Engaging with nonhuman difference through recreational angling. *Environment and Planning D: Society and Space*, 29 (2), 336–352.

Bennett, J., 2001. *The enchantment of modern life: attachments, crossings, and ethics.* Princeton, NJ: Princeton University Press.

Boal, A., 2000. *Theater of the oppressed.* Sidmouth: Pluto Press.

Cahill, A., 2008. Power over, power to, power with: shifting perceptions of power for local economic development in the Philippines. *Asia Pacific Viewpoint*, 49 (3), 294–304.

Cahill, C., 2007. The personal is political: developing new subjectivities through participatory action research. *Gender, Place and Culture*, 14 (3), 267–292.

Carolan, M. S., 2006. Do you see what I see? Examining the epistemic barriers to sustainable agriculture. *Rural Sociology*, 71 (2), 232–260.

Crossley, N., 2007. Researching embodiment by way of 'body techniques'. *The Sociological Review*, 55 (s1), 80–94.

Csikszentmihalyi, M., and Csikszentmihalyi, I. S., 1992. *Optimal experience: psychological studies of flow in consciousness.* Cambridge, UK: Cambridge University Press.

Deleuze, G., and Guattari, F., 1988. *A thousand plateaus: capitalism and schizophrenia.* London: Bloomsbury.

Gieser, T., 2008. Embodiment, emotion and empathy: a phenomenological approach to apprenticeship learning. *Anthropological Theory*, 8 (3), 299–318.

Ginn, F., 2008. Extension, subversion, containment: eco-nationalism and (post) colonial nature in Aotearoa New Zealand. *Transactions of the Institute of British Geographers*, 33 (3), 335–353.

Haraway, D. J., 2008. *When species meet.* Minneapolis, MN: University of Minnesota Press.

Harrison, P., 2009. In the absence of practice. *Environment and Planning D: Society and Space*, 27 (6), 987–1009.

Hitchings, R., 2003. People, plants and performance: on actor network theory and the material pleasures of the private garden. *Social & Cultural Geography*, 4 (1), 99–114.

Hume-Cook, G., Curtis, T., Woods, K., Potaka, J., Tangaroa Wagner, A., and Kindon, S., 2007. Uniting people with place using participatory video in Aotearoa/New Zealand. *In*: S. Kindon, R. Pain, and M. Kesby, eds. *Participatory action research approaches and methods: connecting people, participation and place*. London: Routledge, 160–169.

Ingold, T., 2000. *The perception of the environment: essays on livelihood, dwelling and skill.* London: Routledge.

Ingold, T., 2011. *Being alive: essays on movement, knowledge and description.* London: Routledge.

Jones, O., and Cloke, P., 2002. *Tree cultures: the place of trees and trees in their place.* Oxford: Berg.

Kesby, M., 2005. Retheorizing empowerment-through-participation as a performance in space: beyond tyranny to transformation. *Signs*, 30 (4), 2037–2065.

Kesby, M., Kindon, S., and Pain, R., 2007. Participation as a form of power: retheorising empowerment and spatialising participatory action research. *In*: S. Kindon, R. Pain, and

M. Kesby, eds. *Participatory action research approaches and methods: connecting people, participation and place*. London: Routledge, 19–25.

Krieg, B., and Roberts, L., 2007. Photovoice: insights into marginalisation through a 'community lens' in Saskatchewan, Canada. *In*: S. Kindon, R. Pain, and M. Kesby, eds. *Participatory action research approaches and methods: connecting people, participation and place*. London: Routledge, 150–159.

Krzywoszynska, A., 2012. *'We produce under this sky': making organic wine in a material world*. Thesis (PhD). University of Sheffield.

Krzywoszynska, A., 2015. On being a foreign body in the field, or how reflexivity around translation can take us beyond language. *Area*, 47 (3), 311–318.

Krzywoszynska, A., 2016. What farmers know: experiential knowledge and care in vine growing. *Sociologia Ruralis*, 56 (2), 289–310.

Latour, B., 2004. How to talk about the body? The normative dimension of science studies. *Body & Society*, 10 (2–3), 205–229.

Lea, J., 2009. Becoming skilled: the cultural and corporeal geographies of teaching and learning Thai Yoga massage. *Geoforum*, 40 (3), 465–474.

Lorimer, J., 2007. Nonhuman charisma. *Environment and Planning D: Society and Space*, 25 (5), 911–932.

Lorimer, J., 2008. Counting corncrakes: the affective science of the UK corncrake census. *Social Studies of Science*, 38 (3), 377–405.

Merleau-Ponty, M., and Smith, C., 1996. *Phenomenology of perception*. Delhi: Motilal Banarsidass Publishers.

Morales, M. C., and Harris, L. M., 2014. Using subjectivity and emotion to reconsider participatory natural resource management. *World Development*, 64, 703–712.

O'Connor, E., 2007. Embodied knowledge in glassblowing: the experience of meaning and the struggle towards proficiency. *The Sociological Review*, 55 (s1), 126–141.

Saville, S. J., 2008. Playing with fear: parkour and the mobility of emotion. *Social & Cultural Geography*, 9 (8), 891–914.

Sennett, R., 2009. *The craftsman*. London: Penguin.

Thrift, N., 2008. *Non-representational theory: Spaces, politics, affect*. London: Routledge.

Wijnendaele, B. V., 2014. The politics of emotion in participatory processes of empowerment and change. *Antipode*, 46 (1), 266–282.

9 Shadows, undercurrents and the *Aliveness Machines*

Jon Pigott and Antony Lyons

Introduction: resident in expanded ecologies

The River Torridge catchment (Devon, South West UK) is a picturesque and serene wooded river valley setting, with a largely undeveloped estuary coastal zone. It has been selected as one of six sites in England to be designated a UNESCO Biosphere Reserve. Beneath this seemingly idyllic surface however, there exist some serious ecological-health issues that are being actively addressed by the reserve's operational staff. Key indicators of declining health include the freshwater pearl-mussel, whose habitat lies within the gravel-beds of the river. This shellfish species is rare in the UK, and noted for its longevity and high sensitivity to water quality. An individual's age can be determined by counting the annual growth rings on the shell. In the River Torridge, the local population has not reproduced for over 50 years. Thus, unless the water quality and habitat conditions can become conducive to reproduction again, the concern is that the freshwater mussel will become locally extinct.

This situation is symptomatic of a loss of biodiversity, both within the catchment and more widely, with marked declines also observed in salmon and bat populations.[1] Adding extra complexity is the fact that the life cycles of the salmon and mussel are intimately connected, and both are negatively impacted by the increasing build-up of waterborne silt and mud, with associated high levels of turbidity. These species are bio-indicators; they are the 'canaries in the coalmine' – warning of larger shifts at play and of knock-on effects on wider ecological systems, as well as on the local rural economy through tourism and leisure fisheries. The pervasive damage to ecosystems, such as the Torridge, from agro-chemicals and soil run-off caused by land-use practices can be gradual and cumulative, occurring at landscape, regional and global scales.

Monitoring and data harvesting may reveal specific trends – at least to a technical audience of conservationists and environmental governance agencies. For the most part, however, the chronic accumulation of pollution proceeds beyond the radar of human perception and the need for significant changes to land use and catchment management is difficult to express politically and culturally. *Shadows and Undercurrents* was the name we gave to our rural eco-art project in the catchment of the River Torridge (see Figs. 9.1, 9.2 and 9.3),[2] which sought to respond

Figure 9.1 River Torridge fieldwork.

Attribution: Antony Lyons.

Figure 9.2 River Torridge estuary.

Attribution: Antony Lyons.

Figure 9.3 Ultrasonics workshop with school.
Attribution: Jon Pigott.

to some of these challenges. In the project, we explored methods that may help generate a deeper awareness, empathy and understanding of the co-dependency of the hidden processes and flows. The situation called for a questioning of the status – both material and conceptual – of the many 'actors' involved (Latour 2005). For us, this included contemplation of water as a participant in the mesh of activity.

In this chapter, we draw on a range of theoretical positions including the thinking of PAR scholar Peter Reason, media theorist Jussi Parikka, STS scholar John Law and philosopher Felix Guattari to discuss our work and to explore concepts of water's participation in the blurred world of environmental and media ecologies that the *Aliveness Machines*, (sculptures created as part of *Shadows and Undercurrents*), speak to and for. In the following sections, we first discuss the project in general terms before exploring some of the conceptual frameworks on which we drew. We then outline some of the technical and practical aspects of the creative making process, as situated in an experimental laboratory context. Finally, we reflect on the project overall and particularly how it might be seen as one that invited the participation of nonhumans, via aspects of both practice as research (PaR),[3] and Participatory Action Research (PAR).

The project

As artists-in-residence over a period of 18 months, we engaged with the site, the local communities and with innovative data-gathering techniques, aiming to respond to, and reveal, unseen processes. These creative approaches involved extensive exploratory fieldwork, participatory workshops and sound walks with school groups whereby, through the use of sensitive microphones and hydrophones, the human participants were enabled to extend their sensory awareness into hidden aspects of the environment. The culmination of this 'slow-art residency' project was an immersive, scenographic installation space, assembled around the *Aliveness Machines* – a pair of kinetic sculptural works activated by the 'data' gathered. Through this sculptural animation, emergent sound effects and play of light and shadow, we attempted to amplify the changing levels, and complexity, of what were identified as some key hidden ecological processes in the field area. The installation was exhibited at the end of the project at the Appledore Arts Festival located within the Biosphere Reserve.

The *Shadows and Undercurrents* installation brought together some of our efforts to encounter and communicate hidden aspects of the local ecological mesh, or *umwelt* (von Uexküll 2010).[4] As a practical experiment, we explored a pathway towards a new conception of an integrated ecological-health indicator, and towards a poetic synthesis of vital signs suggestive of the locality's ecological aliveness. From its earliest uses, the word *ecology* has reached beyond biological and environmental sciences, to cultural studies, sociology and politics. Current expanded meanings include 'ecologies of place' (Thrift 1999), and 'ecologies of mind' (Bateson 1972), which entails the acknowledgement of both rational conscious and creative unconscious forms of knowing. There is also the notion of 'media ecologies' (see Parikka 2012, Parikka and Hertz 2012), which addresses our increasingly electronically mediated lifestyles in ways that go beyond the human, extending to the material and even to the geological aspects of media technologies. Parikka suggests that such technologies, often highly refined and deeply integrated into human lifestyles, can be explored in ways other than through their typically screen-based human interface (Parikka 2012, p. 429). Behind their screens, these devices are assemblages of manufactured materials, as well as minerals, such as coltan and gold, which have accumulated over geological time before being folded into human communication and information structures.

The intersections of these expanded notions of ecology informed the creative collaborative project,[5] which was situated within environmental and site-specific art, geopoetics and deep mapping.[6] Although we had a range of objectives, a crucial aspect of the project in relation to this collection was its character as an extended durational investigation of the landscape, through slow attunement and creative 'listening' (see Fig. 9.4). This process involved a distillation of a rhizomic mesh of conversations and encounters, embracing place identity, species, technology and communication. Bat activity from the shadows, and river pollution parameters from the undercurrents, were the key data streams that we chose to work with, revealing aspects of nonhuman realms through the kinetic, sculptural *Aliveness Machines* operating in response to live and recorded data flows. In this way, our experiment was founded on a very limited set of data flows, but also introduced a multi-sensorial information-rich space for embodied human response, or affect. With the *Aliveness Machines* we aimed to raise interest and questions around both

Figure 9.4 In the River Torridge.
Attribution: Antony Lyons.

the ecological vitality of this bioregion *and* the means by which we come to measure and understand it. Our creative collaboration therefore necessarily involved two interwoven perspectives: one emerging from a fusion of 'intimate' environmental sciences and intermedia installation, the other from a concern with tools of measurement explored through the context of kinetic sound art.

Paradoxes and intimacies of knowledge

Working between technologically enabled approaches/methods, and an embodied connection to the 'natural' ecology and environment of the Torridge catchment, exposed a tension. The problem is neatly summarised by Timothy Morton in a lecture titled *This Is Not My Beautiful Biosphere*:

> The dilemma of an ecological era is that the era is at once the product of massively increased knowledge, but also that this knowledge is itself a product of a planetary-scale imagination that has already profoundly damaged the earth.
> (Morton 2012, n.p.)

The 'massively increased knowledge' that Morton describes enables the technological intervention into, and measurement of, the ordinarily less accessible environmental bio-web, thus deepening our understanding of the natural processes and systems at play. It also suggests the contemporary possibility (or fiction) of living at arm's length from 'nature', in contexts ranging from virtual software worlds and 'second-lives' to urban 'bubble' arcades; environments which contribute to conditions such as 'nature deficit disorder' (see Louv 2005). Despite this sense of disconnect, we humans remain multi-sensorial beings, experiencing the intimate, haptic sensation of rain on our skin, or the floodwaters lapping around our feet. However, it is generally thought to be impossible to directly feel climate change, or the slow gradual ecocidal decline in global biodiversity.

Roger Malina (2009a, p. 184) highlights such concerns in his thoughts on 'intimate science' and the 'hard humanities'. Malina considers issues relating to the fact that as a scientist, almost none of the information about the world that he studies is captured by his 'naked senses'. Instead, he develops an understanding of the world through an intimacy with his scientific instruments, inventing new words to describe new phenomena and knowing intuitively when his 'instrument is hallucinating' (2009a, p. 184). Such intimacy, however, is not in the daily experience of most people. He outlines developments that address this issue, including the work of practitioners such as sound-artist David Dunn, who harnesses environmental data flows for cultural purposes, 'coupling the virtual world to the physical, making [data] intimate, sensual and intuitive' (Malina 2009b, n.p.). In this presentation Malina also discusses the role of 'micro science', embedding technological mediation into everyday life creating the possibility of 'open observatories' for local and community knowledge and data-acquisition. An example of this approach, within the context of social science, is the *Morris Justice Project*, set up by members of the Public Science Project in New York City (Stoudt and Torre 2014). Here, participatory data-gathering methods were used to map events and activities regarding the New York Police Departments use of stop and frisk measures. The data gathered by communities around the city was used to communicate issues, raise questions and challenge policy with regard to discriminatory policing and effective and efficient crime prevention. The project shows the potential of the 'open observatory' within a socio-political context. The approach was one we were keen to deploy in the socio-ecological context of the River Torridge catchment.

Our leaning towards an open source, demystifying and participatory approach to instrumentation and data handling nurtures an 'intimacy' with science-based technology and resonates with wider accounts of participatory research, including Reason's model of 'participation as education and transformation', which seeks to challenge powerful groups in society monopolizing 'the production and use of knowledge for their own benefit' (Reason 2005, p. 38). Much of the technology that enabled our project (see Fig. 9.5), particularly in the data-acquisition stages, came from and was supported by an open-source culture. The term 'open source', often associated with computer coding but equally applicable to a broad spectrum of technologies, describes the way in which all information necessary for the use, modification, adaption and application of a technology is freely available. This provides the basis for self-organising communities and groups of participants of all abilities who share tips, information and concepts for the use and application of technologies (see DiBona *et al.* 1999). Such relationships exist in contrast to the hierarchical paradigms of technology corporations guarding 'intellectual property' for profit and market domination.

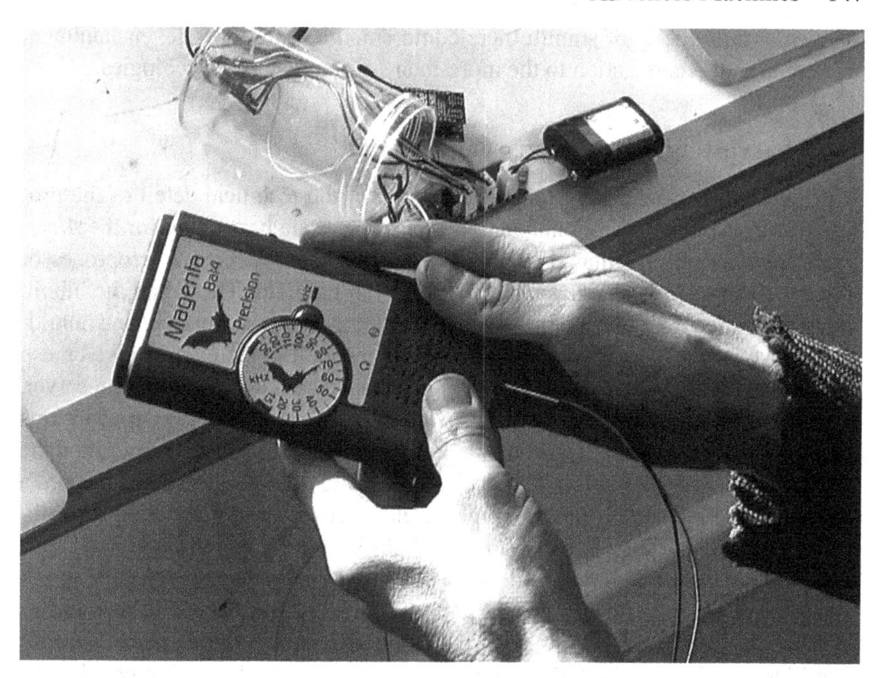

Figure 9.5 Hacking a bat-detector.

Attribution: Antony Lyons.

To further draw out links with participatory research methods, we see our experimental work as fusing creative science and science-based creativity, thus utilising what Reason calls an 'active science', an approach which 'integrates all forms of knowing – immediate acquaintance, aesthetic expression, informative statements, practical competence – in [. . .] inquiry and education process' (Reason 2005, p. 38). The *Aliveness Machines* represent an attempt to make and present something sensual and intuitive from ecological data streams by reflecting notions of open technologies and tools of observation. Within the mix, we integrated objects, materials and mechanisms that could manifest data in ways that are performative, sensuous and atmospheric, and are subjective and personal whilst also being informative. We sought to render detectable what is largely undetectable to our senses, by augmenting technologies and hacking readymade objects into assemblages capable of evoking resonant and richly metaphorical associations (described in more detail below). For us, these were 'provocative prototypes' that tested and demonstrated a proof of concept, encompassing fieldwork and technological sensors, through to gallery installation and audience.[7] At the same time, our assemblages operated as entities that were reflective of the relationship between creativity, technology and ecology. Mediated by digital technology and data, our installation space was scenographic; to produce it, we created a stage with props, robotic actors, projections and soundscapes. There was a mycelial, or rhizomic, strategy in operation; a proposition that through the noise of such a multiplicity of stimuli there could emerge some intrigue, enchantment and emotional reconnection to the more-than-human realms, or ecologies.

Data harvesting, translations, engagement

It is worthwhile now briefly to consider some of the technical detail of the project within the context of nonhuman participation. Data harvesting for the *Aliveness Machines* was facilitated by the placement of small wireless microprocessor devices (using *XBee* and *Arduino* technology) in the field. These had the inbuilt capability to record a range of parameters including light, temperature and humidity. Transmission of periodically sampled values took place over a linked/mesh network to the nearest internet connection, relayed to a dedicated online server, as well as creating an archived log. An initial challenge was how to adapt this technology to monitor the levels of silt pollution (or turbidity) of the river, and the activity of the bat population in the area. Measuring the river turbidity involved a reasonably simple appropriation of the wireless sensor's light-detecting capabilities. The murkier the river water, the less light is able to pass through it. Thus, a light sensor (in this case a light-dependent resistor) makes for an effective monitor. Sensing the more-than-human realm evidenced by the ultrasonic frequencies of bat-calls was achieved by hacking the output of a basic heterodyne bat detector (the Magenta 4) onto the XBee wireless detectors input. This set-up enabled the 35Khz-and-above frequency range of the Pipistrelle bat call and the 80Khz-and-above call of the Horseshoe bat, for example, to be converted into a set of data, representing bat activity that was sensible to humans who are unable to hear anything outside of a 20Khz frequency band.

These approaches to sensing variations in river pollution and detecting bat activity are not particularly new in a monitoring context. Environmental agencies and wildlife groups use similar methods for their sampling and monitoring surveys. What made the project novel, from a technological and participatory point of view, was that the sensing was being achieved through relatively low cost and easily available means, and through the support of an open-source culture of sharing knowledge and resources. Furthermore, the data was live and accessible via the internet, reflecting an open observatory and micro science approach. Through the workshops that we facilitated as part of the project, school groups were able to monitor and work with these data feeds, and become familiar with the technology and its placement in the environment (see Fig. 9.6). For us, the designing, making and testing of elements of the data-sensing technology also provided another perspective on the relationalities involved. Hands-on involvement in the encounters between the technological and environmental realms, and the ways in which these exchanges were translated into data, led us to a deeper acknowledgement of the agency of nonhuman participants within the project.

Mitchell Akiyama (2014) writes that digital technologies, and a general increase in the amount of data available, have influenced the rise in data sonification as creative practice since around the 1960s. He emphasises a growing concern at this time with bureaucracy and the rise of an information society, as

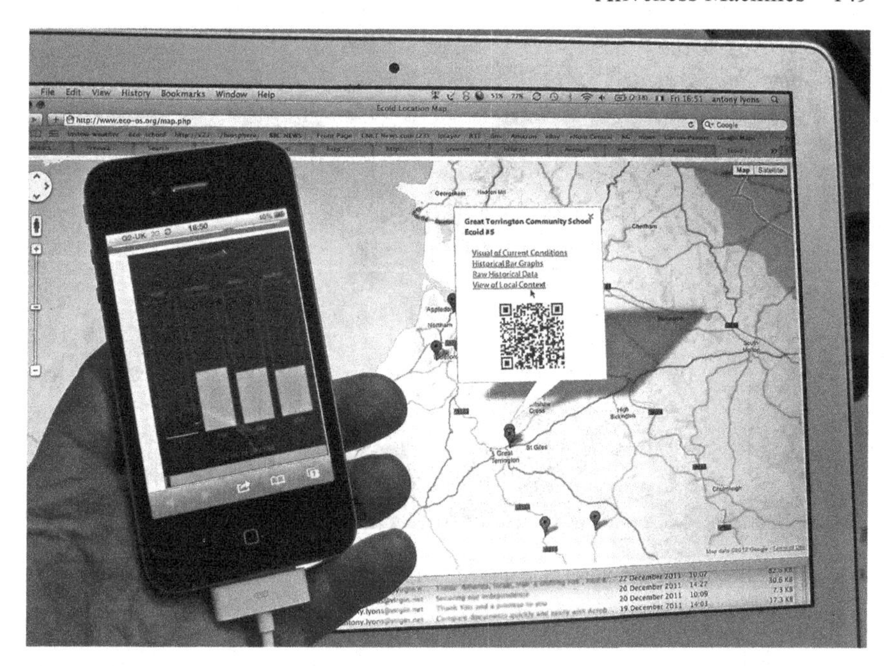

Figure 9.6 Online schools-based sensor data.

Attribution: Antony Lyons.

being concurrent with early examples of this mode of practice. With the modern availability of open source-type programming packages such as Processing and Max/MSP, along with easily available rich site summary (RSS) feeds of data via the internet, as well as other developments such as wireless connectivity, creating digital visualisations and sonifications from data clearly offers a fertile territory for creative practice. Noortje Marres notes, however, that there are potential problems with public engagement through ubiquitous data streams creating a kind of 'informational citizenship' (2012, p. 5). She describes this as a form of public participation that makes impossible demands on people, insisting they take interest in complex issues that have little relevance to their everyday lives (2012, p. 5).

Marres instead develops the idea of 'material participation' as a form of public participation that is enacted with materials and objects and as something that can work alongside and complement 'informational citizenship' (2012, p. 5). She explores this through empirical examples relating to sustainable living with a focus on action and impact rather than just understanding and knowledge. For us, the situated field-experience of the project residency, and the material complexity of the natural environment with which we were working, warranted more than a purely digital, screen-based visualisation or loudspeaker-based sonification of any data taken from the field. Uncritical data visualisation practices have been challenged by Robert Kosara (2010) and others. Embracing complexity and

Figure 9.7 Bat-wheel aliveness machine (detail).
Attribution: Antony Lyons.

paradox, we developed the *Aliveness Machines* to critically and imaginatively intervene with the data flows, whilst exposing the dynamics of some local ecological processes. Through sculptural animation, mechanical sound events, light and shadow, we thus blurred the boundaries between technological and environmental ecologies (see Fig. 9.7).

The aliveness machines

In a productive sense, the project involved the construction of two kinetic sculptural works as centrepieces to the installation space. The first such *Aliveness Machine* was created in response to the activity of selected bat populations in the Biosphere Reserve. This device harks back to early cinematic contraptions such as the *Zoetrope* (whose name means *wheel of life*). Our 'bat-wheel' was a clear acrylic disk with an embedded sequence of silhouettes of a bat in flight (see Fig. 9.8). When rotated by a motor, with a synchronised strobe light, the flickering shadows of the flying bat are projected onto a screen. Another cinematic, audio-visual component was added, akin to a *Mutoscope* (an early cinematic device), which, through its simple flick-book style animation, contributes a further mechanical flapping sound. The spinning of the wheel and associated sounds of the mechanism were triggered by the bat activity recorded and fed from the field via the wireless sensor devices. The exposed mechanisms and technologies of the bat-projecting wheel served to reveal the assemblage nature of the technological mediation that was at work between a living bat in the field and the immersive installation, whilst also allowing for a poetic and human connection to the source of the data. The connection was reinforced by referencing the wings of a flying bat, as well as alluding

Figure 9.8 Bat-wheel aliveness machine (detail).

Attribution: Antony Lyons.

to the chatter of old film projectors, exhibiting another layer of association with silent-era vampire films. This intimacy between technology and atmospherics, data and environment would have been very difficult to achieve through a purely screen-based visualisation output.

The second *Aliveness Machine* was developed to reflect the turbidity/silt levels in the hidden depths of the river. It also used an interplay of light and mechanical sound sources to represent activity and flux in the natural environment. Light was a particularly appropriate medium with which to work, as it was the cloudiness of the water and its propensity to pass light that was being measured by the sensors situated in the field. To pursue this theme a bundle of reflective steel ribbons was suspended inside a cylindrical cage formed from hundreds of fishing lines illuminated from below by a bright light (see Fig. 9.9). When the river was clear, the associated data stream activated a bladeless fan, which in turn excited the steel ribbons causing chaotic patterns of light, reflecting and diffracting on and through the translucent fishing line. Each of the steel bands was also attached to a contact microphone so that when it was excited a mid-frequency rumble became audible, reminiscent of underwater sonics. This water-pollution-focused *Aliveness Machine* was also partnered by a bolt-on, data-activated fishing reel mechanism, which further augmented the soundscape through its characteristic clicking sound. On one of our field outings, we had taken a canoe trip along the river, viewing the landscape from a vantage point usually only enjoyed by the resident wildlife and the occasional fisherman. During this trip, the sound of fishing reels merged seamlessly with the background soundscape of the river. The inclusion of these relational installation elements can be seen as eco-metaphorical. We aimed to echo elements of the river's soundscape as well as poetically reflecting how clear water can equate to good levels of fish stock, thus having a beneficial effect on the local economically important angling activities.

Figure 9.9 Pollution column aliveness machine (detail).

Attribution: Antony Lyons.

Theoretical reflections

The kinetic, optical and sonic nature of the *Aliveness Machines* within the immersive *Shadows and Undercurrents* scenographic installation was, in part, designed to make manifest the material complexity of both the ecology of the Biosphere Reserve's environment and the technological assemblages used to measure and monitor it. Neither of these two complexities, it was felt, were really fully described by, or usefully reduced to simple 'data'. John Law provides some useful understanding in this regard when, in a discussion of what may be considered part of the standard scientific method, he states that 'the materiality of the process gets deleted' (2004, p. 20). Laboratory and field based experiments typically involve a rich interplay of a myriad of animal, mineral, plant and human entities, which are capable of throwing up ever-surprising streams of events. Law's point is that typically much of this interesting activity is subsequently deleted in what is ultimately published in the form of papers, spreadsheets and graphs. Further to this, Law explores the role of the very objects and technologies that are used to measure and monitor the more-than-human world around us. These measurement or 'inscription' devices as Law refers to them (2004, p. 29), so central to the harvesting and extraction of data, are themselves material complexities and assemblages of human and nonhuman agencies. The complex nature of these devices and processes however is typically hidden away, physically obscured inside a case or box, hiding all the inter-dependant relationships of the various components. Law describes the process of obscuring the complex relationships involved in producing data (whether through relying on written numbers and text or through the use of neatly hidden technologies) as a 'hinterland of scientific routinisation' (Law 2004, p. 35) and through the idea of the 'black box' (2004, p. 34), a term with its origins in military systems design.

Similar themes to this are also present in Jussi Parikka's (2012) writings within the field of media archaeology, an area that takes an interest in the material make-up of media technologies. Devices such as computers and smartphones that are readily associated with, and folded into, a textual and cultural domain of human meaning and commerce are, according to Parikka, also part of a nonhuman history of 'media-natures' (2012, p. 97). He suggests that these can be viewed as one big chemical reaction involving glass plates, gutta percha, shellac, silicon, copper and all other manner of minerals and component parts that happen to be available from the natural, material environment. By viewing the machines that we use to communicate, measure and record the world around us in terms of their components and material make-up, media and other technologies can be placed in a geological or geophysical context just as much as a human one. From this perspective, media history reaches back beyond any human history.

Our experimental efforts led us to the realisation that the material complexity of our *Aliveness Machines*, and the methods by which they operated, were important and active participants in the project. The bats and suspended particles in the water also contributed their part to the creative outcome and hence to everything that was ultimately communicated to our audiences. Whilst at an early design stage the *Aliveness Machines* had been conceived of without the input of these nonhuman participants, through iterative development of the project, room was created to – potentially – allow the voices of both our human collaborators and the beyond-human entities to be heard. The appropriated technologies and materials, and the unpredictable and unfamiliar (to us) worlds of bats and water were as equally responsible for the creative outcome as we were. Yet we had not been able to design the project entirely from its inception with these co-producers in the way that a fully participatory project might have aspired to do, by allowing participants to lead and design as well as partake in research (Banks 2012, p. 8). The fact that the artist residency was happening at all was tied to different, quite anthropocentric ecologies of arts and regional development funding, educational upskilling agendas and many planning meetings over coffee etc. However, our creative method, based on slow, deep mapping and careful listening to the landscape, did help us to take our cues from, and reflect on, what we encountered beyond the human realm.

Feminist STS scholar Karen Barad offers some insight that, for us, resonates with the creative processes behind the *Aliveness Machine* elements. In a discussion on the use of piezo-electric transducers and ultrasonography within medical applications, Barad explores the relationship between the 'material and the discursive', something central to her notion of 'agential realism' (2007, p. 191). Ultrasonography, and the mechanism through which ultrasound is transduced and used within a medical context, clearly has interesting parallels to the detection of bats via their use of ultrasound for navigation and hunting. Barad describes similar technologies to those that we deployed in the North Devon Biosphere Reserve as devices for 'making and remaking boundaries [. . .] between human and non human, living and non living, visible and invisible' (2007, p. 201). Other observations by Barad reflect on the interplay between the science, technology and creative practice of the *Shadows and Undercurrents* project, while also reinforcing some of the points made by Law:

> Apparatuses are not pre-existing or fixed entities; they are themselves constituted through particular practices that are perpetually open to rearrangements, re- articulations and other re-workings. This is part of the creativity and difficulty of doing science.
>
> (2007, p. 203)

The practicalities of our creative residency raised our awareness of these theoretical positions as we worked in the field with sensing and measuring technologies designed to operate across widely available platforms of digital media, drawing together a web of environmental and technological interactions (see Fig. 9.10). Issues of river access, fishing seasons, power supplies, dead batteries, mud, wireless ranges, web connectivity and waterproofing were just some of the challenges encountered in any one day. It became clear that the artistic outcome from the project would need to reflect this material complexity and the tenuous nature of the influences that led to what would ultimately become regarded simply as 'data'.

The particular data we were keen to explore, relating to bats and river pollution, emerged from a sequence of countless translations between one material

Figure 9.10 Pollution aliveness machine set-up.

Attribution: Simon Warner.

substrate and another. Ultrasound bouncing between trees and insects became electromagnetic waves, binary numbers, mechanical force and flickering shadow. Clouds of mud and microscopic particles in the riverbed became a factor in natural light levels, a voltage level held on a on a server, the clicking of a fishing reel. Many of these translations were carried out by manufactured socio-technological black boxes such as bat detectors, silicon chips and internet service providers. Further down the line more translations were enacted by our *Aliveness Machines*. Through their assemblage aesthetic of exposed mechanics, localised light sources and sounding materials, the aim is that the *Aliveness Machines* draw attention to the process of translation rather than hide it away behind some kind of screen or inside a sleek black box. In this way, the *Aliveness Machines* are able to raise questions around both the state of ecological vitality *and* the means by which we may come to measure and understand it. Over the course of the project it became clear that the role of the creative visualisation and sonification of environmental data extended across the entire operation of sensor design and placement in the field through to the creation of the interplay of movement, light and sound to describe something of the data stream and its origins within the context of a scenographic immersive installation.

More than human: water as participant

In addressing ideas of expanded ecologies, Guattari's (2000) concept of 'the three ecologies' represents an enduring and overarching context for our creative efforts. His argument was that it is only through broadening our view of the meshes of the ecologies of the environmental, social and mental that we can bring about the necessary shifts away from destabilisation of our planetary life support capabilities, and away from destructive resource-depletion. For example, he writes:

> Wherever we turn, there is the same nagging paradox: on the one hand, the continuous development of new techno scientific means to potentially resolve the dominant ecological issues and restate socially useful activities on the surface of the planet, and, on the other hand the inability of organised social forces and constituted subjective formations to take hold of these resources in order to make them work.
>
> (Guattari 2000, p. 31)

Here, Guattari provided one articulation of some ideas that are becoming increasingly relevant today, especially through transdisciplinary and interdisciplinary research, across fields of socially engaged and participatory art-practice, human geography, place-research, ecological sustainability and health and well-being. Many such approaches can be considered *geopoetic* – embracing diverse contacts with a locality, blending ecological perspectives with poetic imaginaries; entering 'a mental space where conjecture and imaginative play are needful and legitimate' (McKay 2011, p. 10). In framing our exploratory thinking, we drew on a wide breadth of modes of re-imagining ecological relationships and possibilities. In *The Spell of the Sensuous*, anthropologist-magician David Abram speculates that 'despite all the mechanical artifacts that now surround us, the world in which we find ourselves before we set out to calculate and measure is not an inert or mechanical object, but a living field, an open and dynamic landscape subject to its own moods and metamorphoses' (1997, p. 32). Based on research amongst traditional medicine people and shamans, Abram explores the need for 'boundary keepers' who are 'the intermediaries between the human community and the more-than-human community' (2006, n.p.). He asks, 'Look at the river. Do you know how the river feels whenever the salmon returns to its waters?' (2006, n.p.).

In the spirit of this, and as an extension of our *Shadows and Undercurrents* explorations of place, we were invited to co-design an intensive workshop as part of the *In conversation with. . .* project (discussed in more detail by Bastian and Heddon in their respective chapters in this volume). This project looked at the possibility of more-than-human participatory research and the focus for the workshops was participation *with* an element – in our case water. A key text for discussion during the workshop was Ivan Illich's (1985) 'H2O and the Waters of Forgetfulness', amongst others. This work describes the shift away from a holistic relationship with water towards the contemporary prevalent and narrow view of this life-giving substance as simply a utilitarian chemical compound. It prompted us to consider the intrinsic 'aliveness' of water, or a river, or an ecological zone. In challenging anthropocentrism and the division of the world into living and non-living, into 'animal, vegetable or mineral' we began to speculate on how we might 'speak' to a river, or for a river, or even as a river. Our bodies are more than 70% water. It is alive within us. Where then do we draw the line between 'alive' and 'inert'? Like the river, which has no definable boundary, could our bodies also be considered to merge with the 'outer' world in a continuum? Leonardo da Vinci reflects this sentiment when he noted 'as from the said pool of blood proceed the veins which spread their branches through the human body, in just the same manner the ocean fills the body of the earth with an infinite number of veins of water' (in Keele 1983, p. 80).

The group of participating academic researchers set out to examine the co-production of knowledge, expanding beyond the human realm to include the voices, needs and agencies of nonhumans. The workshop's initial aim, or thought experiment, was to explore the possibilities and obstacles of including water as a participant in the research process. However, in this case, it was the whole River Torridge catchment that gradually emerged as the participatory entity. One of the activities was a trip, by small boat, along the estuary and river during which we attempted, with the aid of data from salinity meters, to gauge the hidden, shifting boundary of salt-water and freshwater. How separate are these water 'bodies'? Where does the river end, and the sea begin? In the liminal setting of the tidal estuary zone, does the river lose its identity in meeting the sea? Our exploratory encounters also included the visceral experience of the elusiveness of the waterbody at its boggy headwater regions. Here, we were left with a sense of water as all-pervasive within the hollows, pores and microscopic rivulets of the land, also present in much deeper veins as hydrothermal flows, echoing the nature of water and circulation in living organisms, human and nonhuman. Through a number of such embodied and immersive experiences, including full immersion through swimming, there emerged responses that considered 'the complex webs of relationality between the river, the area's geomorphology and the many human and nonhuman beings that made their life-ways through and with the river' (Bastian 2013).

Conclusion

Whitelaw describes data as 'a set of measurements extracted from the flux of the real [that] are abstract, blank, meaningless' (2008, n.p.). Our sculptural assemblages and immersive installation space sought to bring the real back into play, returning channels of information to material activity and agency. This is reflective of the fact that in the narrative of this project, our starting point was not simply the data, and our aims not just data visualisation or representation. Our creative, situated experience and critical engagement involved wider questions and challenges relating to our sense of place in a field of nature-culture tensions. These included the following: how to both query and translate the (largely) hidden state of health and well-being of the local ecological mesh of human and nonhuman interactions; issues of harnessing creative science and science-based creativity to enhance human connections to ecological processes; and how to explore, through experimentation, the coupling of the virtual world with the physical, attempting to make data flows sensuous, intimate and atmospheric. In attending to these challenges through participatory methods, we developed tactics and technologies with the support of open source communities and made systems that were themselves open source in terms of both their technical assembly and the accessibility of the data that they produced. With the sculptural *Aliveness Machines*, we explored how nonhuman 'technological' participants contributed to the creative translation of some dynamics within the natural environment. These technological arrangements were ultimately intended to communicate the behaviours of bats and water, two vitally important nonhuman participants within the project. Through the involvement of participant groups, we attempted to facilitate a wider community awareness of, and conversation with, the ongoing monitoring and stewardship of the ecological processes in which we are embedded.

Notes

1 For example, in the periodic review of the Biosphere Reserve, the authors state that 'UK populations of the greater horseshoe bat have fallen by 90% in the last 50 years' (2015, p. 47).
2 This was commissioned as part of a wider project called Confluence, whose broad aims were to involve local communities, creative practitioners and novel digital technologies in order to observe, measure and communicate aspects of the natural ecology of the UNESCO Biosphere Reserve.
3 An institutional term for creative research enquiry and critical reflection, leading to a durable record of new knowledge or substantial new insights.
4 In 1909, Jacob von Uexküll theorised that organisms have different '*umwelten*', even while sharing the same environment. *Umwelt* theory states that the mind and the world are inseparable, because it is the mind that interprets the world for the organism. The small fraction of the world that an animal is able to detect is its *umwelt*. The bigger, fuller reality he termed the *umgebung*.
5 Core functional collaborators included the Biosphere Reserve Team, the IDAT unit of University of Plymouth and David Brinicombe, a local enthusiast engaged in bat observation and recording.
6 Deep mapping is described by artist Iain Biggs as aiming 'to challenge the official management of memory that fixes the value and uses of places' (2010, n.p.).
7 It is important to note that the preparatory data-harvesting efforts were aimed at both demonstrating proof-of-concept in sensor design and gathering a limited data-set for demonstration purposes; we were not aiming, under the circumstances, to collect scientifically robust data – though this is certainly one possible future trajectory for our explorations.

References

Abram, D., 1997. *The spell of the sensuous: perception and language in a more-than-human world.* New York: Vintage Books.

Abram, D., 2006. The ecology of magic: an interview with David Abram [online]. Available from: www.scottlondon.com/interviews/abram.html [Accessed 16 May 2016].

Akiyama, M., 2014. Dataffect: numerical epistemology and the art of data sonification. *Leonardo Music Journal*, 24, 29–32.

Banks, S., 2012. *Community based participatory research: a guide to ethical principles and practice* [online]. Durham: Durham University. Available from: https://www.dur.ac.uk/resources/beacon/CBPREthicsGuidewebNovember20121.pdf [Accessed 16 May 2016].

Barad, K., 2007. *Meeting the universe halfway: Quantum physics and the entanglement of matter and meaning.* Durham, NC: Duke University Press.

Bastian, M., 2013. *Conversations with the elements* [online]. Available from: http://www.morethanhumanresearch.com/conversations-with-the-elements.html [Accessed 19 May 2016].

Bateson, G., 1972. *Steps to an ecology of mind: collected essays in anthropology, psychiatry, evolution, and epistemology*. Chicago: University of Chicago Press.

Biggs, I., 2010. Deep mappings/spectral traces: a partial view [unpublished]. Keynote paper for the "Mapping Spectral Traces" symposium, Virginia Tech, USA, 13 October.

DiBona, C., and Ockman, S., eds., 1999. *Open sources: voices from the open source revolution*. Sebastopol, CA: O'Reilly.

Guattari, F., 2000. *The three ecologies*. London and New York: Continuum.

Illich, I., 1985. *H2O and the waters of forgetfulness*. Dallas: Dallas Institute of Humanities & Culture.

Keele, K., 1983. *Leonardo DaVinci's elements of the science of man*. New York and London: Academic Press Inc.

Kosara, R., 2010. *The visualization cargo cult* [online]. Available from: https://eagereyes.org/criticism/the-visualization-cargo-cult [Accessed 15 May 2016].

Latour, B., 2005. *Reassembling the social: an introduction to actor-network-theory*. Oxford: Oxford University Press.

Law, J., 2004. *After method: mess in social science research*. New York: Routledge.

Louv, R., 2005. *Last child in the woods: saving our children from nature-deficit disorder*. Chapel Hill, NC: Algonquin Books of Chapel Hill.

Malina, R. F., 2009a. Intimate science and hard humanities. *Leonardo*, 42 (3), 184.

Malina, R. F., 2009b. *Making science intimate: the scientific method as a territory for artistic experimentation* [online]. Available from: http://www.diatrope.com/rfm/docs/Imagen_09.pdf [Accessed 15 May 2016].

Marres, N., 2012. *Material participation: technology, the environment and everyday publics*. Basingstoke: Palgrave Macmillan.

McKay, D., 2011. *The shell of the tortoise: four essays & an assemblage*. Kentville, NS: Gaspereau Press.

Morton, T., 2012. *This is not my beautiful biosphere (video)* [online]. Available from: http://ecologywithoutnature.blogspot.co.uk/2012/10/this-is-not-my-beautiful-biosphere-video.html [Accessed 20 September 2015].

North *Devon biosphere review*, 2015. [online]. Available from http://www.northdevonbiosphere.org.uk/review-report.html [Accessed 18 May 2016].

Parikka, J., 2012. New materialism as media theory: media natures and dirty matter. *Communication and Critical/Cultural Studies*, 9 (1), 95–100.

Parikka, J., and Hertz, G., 2012. Zombie media: circuit bending media archaeology into an art method. *Leonardo*, 45 (5), 424–430.

Reason, P., 2005. Living as part of the whole: the implications of participation. *Journal of Curriculum and Pedagogy*, 2 (2), 35–41.

Stoudt, B. G., and Torre, M. E., 2014. The Morris justice project: participatory action research [online]. *Sage cases in methodology*, London: Sage. Available from: http://dx.doi.org/10.4135/978144627305014535358 [Accessed 18 May 2016].

Thrift, N., 1999. Steps to an ecology of place. *In:* D. Massey, J. Allen, and P. Sarre, eds., *Human geography today*, Cambridge: Polity Press, 295–323.

von Uexküll, J., 2010. *A foray into the worlds of animals and humans*. Minneapolis, MN and London: University of Minnesota Press.

Whitelaw, M., 2008. Art against information: case studies in data practice. *The Fibreculture Journal* [online], 11. Available from: http://eleven.fibreculturejournal.org/fcj-067-art-against-information-case-studies-in-data-practice/ [Accessed 31 March 2015].

Part III

Cautions

10 Laboratory beagles and affective co-productions of knowledge

Eva Giraud and Gregory Hollin

> The Beagle's excellent disposition and gay personality are two [of] its greatest assets, because special handling is seldom necessary and a minimum amount of restraint is required for most experimental procedures.
>
> (Anderson 1970, p. 4)

The above account of beagles' value to experimental science comes from a researcher based at the first large-scale experimental beagle colony at the University of California, Davis (1951–1986). The quotation is indicative of the messy combination of cultural and scientific factors that led to the animals' consolidation as the standard laboratory dog and makes evident that the animals' affective qualities – their 'excellent disposition' and 'gay personality' – lie at the core of their experimental value. The tight link between affect and epistemology has been noted in previous research about laboratory dogs (e.g. Dror 1999, Degeling 2008), with dogs in general focused on in a large body of cultural research, due to drawing together issues including the historical standardization of laboratory animals (Kirk 2010); connections between veterinary and medical research (Bresalier *et al.* 2015); and ethical debates surrounding animal research (Lederer 1992). These latter concerns are brought to the fore when focusing on beagles specifically, due to the breed's affective qualities being so closely aligned with beagles' consolidation as the standard laboratory dog. Davis, moreover, is a privileged site through which to explore the standardization of beagles; although the breed had been used in research prior to the 1950s, the experiments at Davis consolidated their use (Thompson 1989). As researchers involved with the project noted: 'the many arguments that can now be advanced for the use of this animal were unknown, or at least unsupported, when the decision to employ beagles in these experiments was made in 1950' (Thompson 1989, p. 25). In addition to the breed's specific qualities, the scale of these experiments at Davis resulted in the 'continued use of the beagle in subsequent experiments' due to the need for 'intercomparing data in the same animal model' (Thompson 1989, p. 25). Research generated at Davis continues to inform the contemporary management of laboratory beagles (Tomkins *et al.* 2011), with beagles acting as the standard dog for use in laboratory work in a range of contexts (Joint Working Group on Refinement 2004), and contemporary licensing standards (which prescribe the amount of space and levels of social interaction required for dogs) based on the needs of beagles (e.g. EU 2010).

In this chapter, we argue that the beagle's reputation as being an amenable research subject is striking, first, because of the way beagles' affective qualities become tied to economic considerations around the cost of research, and, second, because it is theoretically informative. The use of dogs in general, and beagles in particular, within laboratory work, elucidates tensions between more-than-human approaches and participatory research, tensions which hinge on these perspectives' contrasting understanding of what it means to 'co-produce' knowledge (as Bastian *et al.* argue in the introduction). We suggest, more specifically, that – in illustrating the vulnerability of affect to instrumentalization, and its role in molding compliant research subjects – beagles raise questions about the way certain more-than-human approaches have depicted affective human–animal relations as generating ethical responsibility towards specific animals through situated affective encounters (e.g. Despret 2004, 2013, Haraway 2008).

At Davis, the dogs' behavior and, indeed, their personalities demonstrably shaped the research space, the care-taking practices that were employed and even the choice of personnel at the site (Giraud and Hollin 2016). In line with understandings of participation and nonhuman agency that are dominant within more-than-human geographies (Hinchliffe *et al.* 2005, Whatmore 2006, Anderson and Harrison 2010, Braun and Whatmore 2011), or science and technology studies (Jasanoff 2004, Harbers 2005, Pickersgill 2012), knowledge generated at Davis could thus be seen as co-produced by researchers, care-takers, beagles, spatial arrangements – and a host of other actors – in the sense of being co-shaped by an assemblage of agencies that are irreducibly entangled. From a participatory research perspective, in contrast, the role of beagle agency would not necessarily be perceived as co-production. Within socio-economic theory (Ostrom 1990), radical geographic contexts (Pickerill and Chatterton 2006), or social movement studies (Borda 2001), for instance, understandings of 'co-production' are closely linked to a social justice agenda that sees the aim of participation as being to improve the quality of life of those involved (Ostrom 1996). Participatory approaches to research, moreover, entail research partners having the potential to radically re-shape the production of knowledge to suit their own needs (Chatterton and Pickerill 2010). As suggested by the opening quotation, at Davis – in contrast – while the beagles did have agency in shaping the research process, this agency was molded in ways that ensured the animals did not ultimately disrupt pre-determined experimental goals, and foreclosed alternative ethical or epistemological outcomes. As we will discuss later, this operated at both the level of individual beagles and at the species level where breeding has selected for docility and amenability.

In order to explore some of the tensions surrounding the ethical potential of affect, this chapter takes a lead from recent calls to focus on the longer histories and wider contexts of contemporary relations in the laboratory (e.g. Johnson 2015). We adopt a socio-historical perspective to explore the participatory potentials that were created – and undermined – with the consolidation of beagles as standardized laboratory dogs during the mid-20th century. Key Anglo-American examples of canine breed-selection, care-taking developments, and colony-maintenance, which contributed to beagles' eventual standardization, are drawn on to illustrate the ambivalent role of affect in affording nonhumans a more participatory role in the research process. Whilst a range of important moments in the breeding of experimental dogs during the first half of the 20th century are used to establish some general context, in terms of primary materials, our focus is on scientific papers and reflections generated by researchers at the first experimental beagle colony at Davis. Before turning to beagles directly, however, it is necessary to flesh out the existing relationships between participation, more-than-human research, and affect.

Affect and the ethics of participation in laboratory work

Recent debates about affect can be contextualized as part of broader research within multispecies geographies, which has begun to ask how to reconcile questions of social justice with more-than-human frameworks (e.g. Collard and Gillespie 2015). These issues are brought into focus within this text around the more specific question of how to relate ethical approaches more commonly associated with participatory research to more-than-human contexts (see Bastian *et al.*'s introduction to this volume). Various mechanisms have been (or at least can be) introduced to afford human publics a more active participatory role in political contexts, such as consensus decision-making (e.g. Cornwell 2011), or involving communities in co-producing infrastructures that affect their everyday lives (Ostrom 1996). Similar mechanisms have been explored in relation to the production of scientific knowledge, from consensus-conference models that give publics an opportunity to debate the direction of laboratory science (Haraway 1997), to the co-management of natural resources (Berkes 2009). When seeking to engage with nonhumans, however, many of these mechanisms are seen as un-workable due to their inability to participate through these formal processes.

Affect has been a pivotal means of overcoming problems of communicative capacity, as affect is believed to open space for alternative, non-linguistic, modes of communication between species (Lorimer 2007, Greenhough 2014). This communication has been described as 'anthropo-zoo-genesis' (Despret 2004) or 'affective attunement' (Willett 2014), and is seen to be grounded in compassion that is generated through 'corporality' (Acampora 2006) and 'somatic sensibilities' (Greenhough and Roe 2011). Without eliding important distinctions between these perspectives, what these accounts share is the argument that affective bodily relations with animals – often those emerging through everyday care-taking practices and interactions – create space for animals to assume a more active role in the production of knowledge.

As with other key definitions of affect (e.g. Lee 2008), Jamie Lorimer understands affect as a quality of actors that designates the organism's capacity to affect others and be affected in turn (Lorimer 2007, p. 915). While not prescriptive in the relations it fosters, affect is believed to encourage and sustain particular forms of relating. So, for example, if this understanding of affect is applied to beagles, the beagle's 'charismatic' characteristics (e.g. their non-aggressive traits and enjoyment of human attention) could be seen as affecting researchers in various ways that may lead them to care for the animals and celebrate their 'gay personality'; care-taking behaviors which – in turn – may enable the beagles to respond in a 'well disposed' way to researchers' own affects.

Importantly, and while acknowledging that different affects might facilitate different modes of relating (not all of which are congenial as that between humans and beagles), a number of thinkers, again including Lorimer (2015, p. 25), have pointed to affect's ethical potential; openness to being 'affected' can create opportunities to move beyond relationships that are simply co-shaping, to embed emotional engagement and ethical responsibility into these relationships. As Matei Candea notes when relating Stengers and Despret to multispecies contexts, if space is created for affective encounters within the laboratory then this is not only thought to open up vectors of communication between researchers and research partners that enable animals to signify their needs (e.g. Despret 2004), but to also allow these partners to 'object' to certain practices (Candea 2013, p. 108). Affect can, in other words, foster continuous ethical obligations towards research partners (Haraway 2008, Greenhough and Roe 2011). This intersects with the emphasis within participatory research on the need for continuous ethical responsibility to individuals implicated in the research process; a responsibility not satisfied by completing an official ethics form or simply adhering to standard ethical procedures (Banks *et al.* 2013).

The participatory potential of affect, and its capacity to sustain ethical obligations, however, is troubled when examining the emergence of the experimental dog and subsequent consolidation of beagles as *the* experimental breed. By focusing upon the longer histories of experimental dogs, we can see that affective dispositions of sentient beings are open to systematic manipulation. Knowledge gained with experimental dogs may be co-produced, in the sense of being entangled and co-shaped, but may still foreclose ethical and epistemic opportunities. Elsewhere (Giraud and Hollin 2016) we consider this issue in relation to experimental procedures; here, however, we focus on the process of selecting animals for experimental research, due to the profound implications selection processes have for participatory relations in the present.

Why dogs?

Discussions around the workings of affect are helpful in elucidating why the use of dogs in experimental research became so widespread from the late 19th century onwards. Attention to bodily relations played a key role in the use of dogs as experimental research subjects within late 19th- and early 20th-century laboratory research, but dogs also illustrate the ambivalent function of affect in this process. To give a brief overview of the evolution of canine research, by the late

19th century, dogs were being used as models for human disease due to a range of physiological, practical, and affective factors. As canine scientist J. P. Scott describes:

> The dog has long been a favorite animal in medical research, partly because of its size and docility but also because of the availability of large numbers of stray and unwanted dogs at low cost.
>
> (Scott 1970, p. 723)

This emphasis on a particular affective disposition an animal may hold, namely here the 'docile' dog, alongside other – more 'mundane' – factors, is present not just in narratives of research scientists but is reiterated within historical analyses of early (and often unsuccessful) experiments with blood transfusion in Britain and North America. Dogs were not solely used due to their physiological affinities with humans but because of their affective – and hence their communicative – capacities:

> canines were often favoured because they were easy to obtain, relatively easy to handle, and through their expressions and postures their behaviour was easily 'read.' As many pet owners could confirm, their dogs were able to communicate to humans a sense of their physical and emotional wellbeing.
>
> (Degeling 2008, p. 25)

Echoing Despret and Haraway, trans-species communication derived from affective relations was seen as critical in enabling care-takers and researchers to interpret animal behavior and adjust the experiment accordingly. In Otniel Dror's analysis of physiology in this period he, accordingly, argues that attention to well-being was not simply an ethical concern, but an experimental one. Dror, citing the work of physiologist Moritz Schiff, contends that animal emotion had to be managed to ensure that results were standardized, as distressed animals produced experimental anomalies: 'The eradication of pain was not "merely an optional noble gesture" but "aided correct scientific observations" ', 'Physiological knowledge', in other words, 'demanded pain-free animals' (Dror 1999, p. 210).

In the early 20th century, the management of emotion was again emphasized in Anglo-American physiology, but this time it was not due to dogs' capacities to be 'read' by experimenters. Instead – and foreshadowing the ultimate decision to focus on beagles – the emphasis had shifted to the value of dogs' own affective qualities:

> The very qualities that endeared dogs to humans made them vulnerable to researchers [. . .] dogs, in light of their tractable nature, were used in the most extreme experiments, which often involved considerable pain.
>
> (Lederer 1992, p. 64)

The acknowledgement of dogs' affective qualities, and the potential for these qualities to give rise to productive relations in the laboratory, had intensified by the 1950s. This intensification occurred in relation to psychological experiments, where dogs became the focus of work to determine whether environmental factors could have a detrimental psychological impact (Kirk 2014).

Dogs thus illustrate the significance of specific affective capacities in decisions to select a particular species for laboratory work, because their capacity to form bonds with humans was at the heart of initial decisions to use dogs in experimental research. While affect played a pivotal role in facilitating trans-species communication within canine research, the instrumental nature of this communication – its role in easing experimental progress, rather than re-shaping pre-determined goals – means that dogs trouble any easy connection that might be made between affective relations and more participatory forms of co-production. Returning to our case study at the University of California Davis, this troubled relationship is particularly evident in the decision to use dogs, and specifically beagles.

The dogs of Davis

A closer look at the rationale behind the first large-scale beagle colony, at the Radiobiology Laboratory in Davis, helps to elucidate this troubling role of affect. As touched on above, the laboratory was funded through the Manhattan Project in order to study the long-term effects of exposure to various forms of radiation. As is noted by Davis researcher Douglas McKelvie and colleagues, the experimental demands ensured that a very particular type of animal was required:

> an animal with a prolonged life-span was necessary. This requirement eliminated such animals as the mouse, rat, and guinea pig. In addition, the physiological and anatomical features of these animals are not closely related to those of man. The natural choice, some species of nonhuman primate, was ruled out by high cost and difficulties in procurement. The final decision was to use the dog, since it was readily available, easy to handle, adapted to laboratory environment, and was especially responsive to human care.
>
> (McKelvie *et al.* 1971, p. 263)

In this extract, it is immediately noticeable how affect and cost-effectiveness are treated more-or-less synonymously as factors to be considered and controlled; the fact that dogs are cheap *and* the fact that they are responsive to human care are both taken into consideration and are believed to make the dog a valuable tool for scientific research. The affective qualities of dogs, moreover, were touched on by all of the key researchers at Davis (e.g. Andersen 1970), and in broader research literatures that stress their 'social relationship with man' and 'docility' (Scott 1970, p. 723), suggesting these qualities make dogs less intimidating to handle than other research animals under consideration such as calves, sheep and pigs (Zinn 1968, pp. 1884–1885).

In addition its use of dogs, the colony at Davis also needs to be contextualized in relation to a broader push to standardize laboratory animals, which came to the fore by the early 20th century (Kirk 2010). In relation to dogs specifically, researchers' had indicated discontent with the use of 'random source' dogs (e.g. Zinn 1968, p. 1883) because:

> The 'normal' [i.e. 'available'] dog could be severely anemic, infested with fleas, lice, ticks, and intestinal parasites such as amoebae. He could have struggled to survive in a state of malnutrition in a poor neighbourhood, without the care and attention necessary for normal growth and development. He may be influenced by an extreme sense of insecurity and anxiety, if such psychic states exist in dogs – who knows? Even more, consider the possible psychologic trauma produced by his captivity, transportation to the laboratory, neglect, and nonsympathetic care during his imprisonment. His sole visitor was the disinterested caretaker who handled the dog roughly in response to the call of the investigator for a 'normal dog' for today's 'crucial' experiment. . . . Normalcy should be supported by criteria of care and health in dogs as well as in man regardless of the demands of effort and funds. Treat not the dog like a dog but more like a man, or the experimental results will 'go to the dogs'.
>
> (Burch 1959, p. 805–806)

This evocation of the emotional state of the 'random source' dog seems to be advocating precisely the mode of situated attention and care towards individual animals, which has been called for in theoretical contexts. In this instance, however, a concern for individual animals is tightly bound up with epistemological concerns, just as laboratory 'captivity', 'nonsympathetic care', and 'disinterest' were seen as exacerbating the dog's state of anxiety, 'care' and 'good health' were seen as integral to ameliorating these problems and hence to creating meaningful experimental outcomes. The push to standardize results through standardizing dogs, as demonstrated at Davis, thus reflected the need for a steady supply of animals with an equally steady temperament and, as Burch notes, this could only be achieved through carefully managing the affective responses of the animals as well as their breeding.

Standardizing beagles

Given that, for a variety of reasons, so few breeds met the requirements of the laboratory (Andersen 1970, pp. 3–4), in the mid-20th century serious consideration was given to developing a new breed of dog specifically for research purposes (Zinn 1968, p. 1886). Indeed, attempts to develop such a dog appear to have been made in Oregon (McKelvie *et al.* 1971, p. 281). Nonetheless, the beagle quickly became established as *the* standardized laboratory dog for it had a vast number of characteristics it had in its favor (to expand on the opening quotation):

> The most desirable qualities of the Beagle as an experimental dog are its medium size, moderate length of hair coat in two or more colors, even temperament, adaptability to living in groups, representative conformation of the dog, and the lack of need for cosmetic surgery. The Beagle's excellent disposition and gay personality are two of its greatest assets, because special handling is seldom necessary and a minimum amount of restraint is required for most experimental procedures. Its excellent disposition is the result of culling ill-tempered dogs throughout the history of the breed. Although a wide range of behavior traits can be identified in the Beagle, they rarely show aggressiveness, timidity, or shyness.
>
> (Andersen 1970, p. 4)

As with the decision to focus on dogs in general, therefore, the beagle's affective qualities are suggestive of how they are also linked to decisions around the economic rationales for using a particular dog breed. Because 'special handling' is rarely needed with the beagle and because they do not need to be 'restrained' (and pictures of the veterinarians at work at Davis [e.g. McKelvie and Andersen 1966, p. 32] show research being conducted without so much as a lead); the beagle's gay personality actually makes the experiment cheaper to run *and* makes the pre-established goals of the experiment easier to achieve (Giraud and Hollin 2016). Once again, it is worth noting that this is not a one-off claim. The same desirable characteristics of the beagle are stressed repeatedly both by researchers from Davis (e.g. Andersen and Goldman 1960, p. 129) and elsewhere, who stress their 'temperament' (Zinn 1968, p. 1885) and 'extreme degree of nonaggressiveness' (Scott 1970, p. 723).

Despite certain hopes for the role of affect, here we see that it does not, however, afford the beagles' agency within the production of knowledge as would be demanded from the perspective of participatory research. As made explicit in Anderson's characterization of beagles' 'gay personality', for instance, the breed was specifically selected because the animals' temperament made them less likely to resist experimental procedures and disrupt the experiment. This temperament, moreover, was actively constructed through culling 'ill-tempered' animals; what results, therefore, is an animal that is conducive to laboratory work. The barriers to giving beagles greater agency in the research project, therefore, are bound up in their embodied biological histories and, thus, cannot easily be resolved through creating the space to learn how the animals signify resistance, in the way that Haraway and Despret suggest, an issue taken up in more depth below.

This is not to say that affective relations with beagles give no scope for animals to shape the production of knowledge, in the sense of co-shaping discussed above; as we discuss elsewhere (Giraud and Hollin 2016), at Davis the spatial environment of the colony was shaped through knowledge gained via affective relations. This knowledge led to the development of new cage designs and care-taking practices (Andersen and Hart 1955, Andersen and Goldman 1960, Andersen 1964, Solarz 1970), which were designed to maximize animal happiness. Even though cage design was – seemingly – co-produced, however, the ultimate aim of these re-designs was to ensure the dogs' on-going compliance in the experiments. This research, therefore, was decisively 'un-cosmopolitical' in Stengers's sense (2005, 2010, 2011) because a pre-determined experimental goal had already been decided and – though knowledge gained from affective relations shaped the way this goal was achieved – it did not shape the end outcome of the experiments.

The problems beagle research pose to extending questions of participation beyond the human are brought into focus if they are related to participatory research more explicitly. As Sherry Arnstein (1969) notes, participation can take a number of forms: from maximum levels of 'citizen power' (such as delegating substantial power or passing all control to citizens and research partners), to more 'tokenistic' gestures (that include consultation). She warns, moreover, that certain processes can serve a placating purpose by giving the illusion of participation whilst actually being a form of 'non-participation' that goes 'through the empty ritual of participation' without affording others 'the real power needed to affect the outcome of the process' (1969, p. 216). Understood in Arnstein's terms, although the Davis beagles had a degree of agency, the re-aligning of this agency to meet existing experimental goals and to prevent the beagles from disrupting intended 'outcomes' could be seen as precisely the sort of 'manipulation' that she equates to 'non-participation'. With beagles, moreover, it is especially difficult to move beyond 'manipulative' forms of 'non-participation', not just because of power relations within the experimental space itself, but because of longer breed histories, which actively discourage any forms of behavior that do not signify 'consent'.

Beagles, participation, and co-production

The role of beagles in laboratory science, therefore, does not only raise questions about the 'levels' of participation that can be afforded to experimental animals, but foregrounds questions surrounding resistance and consent. Concerns about the power dynamics of experimental research have already been raised within more-than-human approaches. Matei Candea, for example, argues that the difficulty in creating space for animals to signify their needs, especially in the context of laboratory science, arises because it is hard to create space for animals to 'object' to the 'impositions of experimental obligations' and 'resist the authority of science' (2013, p. 109). These concerns assume a more profound significance through the lens of participatory research, which emphasizes the importance of allowing research partners to withdraw consent at any stage in the research process, even if formal procedures were adhered to throughout (Banks *et al.* 2013). If animals have been selectively bred to eliminate certain affective qualities or – on an individual level – had their affective responses systematically manipulated with the experimental context, then this could foreclose future opportunities to signify 'objection'.

In beagle research, the broad difficulties associated with consent are compounded, because the affective relations that could – potentially – be a route into understanding when consent is being withdrawn (Haraway 2008, Greenhough and Roe 2011, Despret 2013) actually become a barrier to the participatory co-production of knowledge. The animals' amenability, coupled with the dynamics of the research process, makes them unlikely to 'object' to what is happening to them even if – technically speaking – space is provided for them to do so. At Davis, for instance, the main signs of beagle discontentment were perceived to be 'digging', 'pacing', and fence-jumping' (Andersen and Goldman 1960, pp. 129–130), and a large body of research (gathered in Andersen 1970) was developed on how to engage with the animals' affective qualities in ways that eliminated these activities in the future. This process of discouraging disruptive behaviors was successful, with researchers eventually being able to handle the animals in routine activities entirely without restraints (Giraud and Hollin 2016, p. 12). Davis is not, moreover, the exception, with contemporary research continuing to take a close interest in beagle 'body language' and experiment with practices which could produce more 'consistent and meaningful' data through eliminating responses such as 'shivering, urination or defecation, and panting' (e.g. Döring *et al.* 2016, pp. 18, 21). The specific embodied histories that underpin beagles' consolidation as experimental dogs, and on-going manipulation of these responses on an individual basis within specific experimental contexts therefore, significantly complicates the potential for affect to foster 'mutually beneficial' outcomes, in the manner intended by participatory research (e.g. Banks *et al.* 2014).

As Thom van Dooren notes (2014, pp. 101–108) it is vital to pay attention to the longer histories and contexts that frame – even seemingly convivial – affective encounters. In Despret's influential account of how trans-species communication can occur through bodily engagement, for instance, she draws on the work of ethologist Konrad Lorenz (2004, pp. 128–132). Whilst agreeing with Despret's analysis of Lorenz's work with geese as being 'grounded in relationships of care that enabled the formation of new kinds of knowledge', van Dooren suggests that 'his technique, while good for learning, may not have been so good for geese' (2014, p. 105). Lorenz's method of deliberately imprinting birds so that they saw him as their primary caregiver, van Dooren argues, produces a '*captive* form of life' that produces 'a lifelong attachment' to humans at the expense of relationships with other members of its species (2014, p. 103, italics in original). These arguments are both pertinent to and complicated by laboratory beagles, who highlight the importance of paying attention to the longer embodied histories that frame particular affective encounters, at the level of the breed as well as the individual. Without paying attention to the constitutive relations that frame affective encounters, there is a danger that a lack of substantive 'objection' could be used as evidence of the lack of coercion involved in experimental contexts, or even a sign of care as with Despret, in a manner that elides any need to reflect still further on experimental ethics. This danger has been present throughout the contemporary history of canine experimentation, as we have shown. In the first decade of the 20th century, moreover, researchers' and care-takers' affective work in ensuring animal 'happiness' was used to deflect anti-vivisectionist criticism, and drawn on as evidence for the animals' well-being. As Dror argues:

> Cannon's [1909] code of regulations governing laboratory procedures involving animals, for example, was written explicitly with the antivivisectionists in mind [. . .] Like many of his contemporaries, he adopted the approach of the late nineteenth-century physiologists who repeatedly emphasized their humanitarian concerns and their use of anaesthetics when confronted by anti-vivisectionists' charges, downplaying the physiological rationale behind their particular concerns with suffering.
>
> (Dror 1999, p. 235)

This logic continued into the mid-20th century, with researchers at Davis themselves acknowledging that guided tours of the facility were designed to illustrate the dogs' well-being to the public. The Veterinary School's annual report, for instance, describes how 'several hundred people visit the colony annually and lecturing on kennel activities continue. An open-door policy has averted public criticism by those opposed to the use of dogs for research' (School of Veterinary Medicine 1961, p. i). In pointing to the level of care given to animals, researchers were able to mask the ultimately instrumental function of affect in ensuring animal distress did not disrupt the experiment. Affective relations, therefore, were not just pivotal to the selection and on-going care of dogs, in ways that ensured smooth experimental progress, but were used to diffuse criticism from anti-vivisectionists for whom dogs had been a potent weapon in gaining public sympathy since the 19th century (again in campaigns within both British and North American contexts and United States, see French 1975, Elston 1987, Lederer 1987).

Conclusion

Beagle research, therefore, poses conceptual and political questions about how to foster dialogue between participatory research and more-than-human approaches, whilst problematizing the potential for affect to facilitate this dialogue. As van Dooren notes, there can be a distinct violence in affective encounters that are portrayed as mutually beneficial; as with his critique of certain practices within avian research, some of the processes of human–beagle engagement that occurred both at Davis and within longer histories of dog breeding took 'advantage of an ontological openness' (in this instance the pliable temperament of the beagles) 'to produce an altered way of life' (2014, p. 102). To echo van Dooren, an interrogation of particular sets of relations involved in beagle research is not to deny that species are entangled, or inevitably co-shape one another, it is – however – intended to foreground how certain, mutually affective, encounters might occur 'at the expense of a whole set of other ways of being' (2014, p. 103). Any affective encounter is contingent on an assemblage of environmental, contextual, and historical factors that can support certain affects (those that ensure the beagle is a compliant research subject, for instance) and foreclose others (such as beagle boisterousness).

While affect might be a fertile ground for trans-species communication (Despret 2004, Roe and Greenhough 2014) or even care (Haraway 2008, Davies 2012), further questions need to be asked about the limitations of these processes and the potential for affect to be used for manipulative as well as participatory ends. Beagles, more broadly, raise urgent questions about whether it is possible to depict animals as co-partners – in the sense intended by participatory research – if their longer breed histories, their spaces of encounter, and who they engage with foreclose the potential to withdraw consent (Banks *et al.* 2013) or go beyond having a limited or tokenistic influence (Arnstein 1969), in order to shape the research in mutually beneficial ways.

References

Acampora, R. R., 2006. *Corporeal compassion.* Pittsburgh, PA: University of Pittsburgh Press.

Andersen, A. C., 1964. Air conditioned cages designed to minimize kennel problems. *Laboratory Animal Care*, 14 (4), 292–303.

Andersen, A. C., 1970. *The beagle as an experimental dog.* Ames, IA: The Iowa State University Press.

Andersen, A. C., and Goldman, M., 1960. An evaluation of an outdoor kennel for dogs. *Journal of the American Veterinary Medical Association*, 137 (2), 129–135.

Andersen, A. C., and Hart, G., 1955. Kennel construction and management in relation to longevity studies in the dog. *Journal of the American Veterinary Medical Association*, 126 (938), 366–373.

Anderson, B., and Harrison, P., eds., 2010. *Taking place: non-representational theories and geography.* Farnham: Ashgate.

Arnstein, S. R., 1969. A ladder of citizen participation. *Journal of the American Institute of Planners*, 35 (4), 216–224.

Banks, S., Armstrong, A., Carter, K., Graham, H., Hayward, P., Henry, A., Holland, T., Holmes, C., Lee, A., McNulty, A., Moore, N., Nayling, A.S., and Strachan, A., 2013. Everyday ethics in community-based participatory research. *Contemporary Social Science*, 8 (3), 263–277.

Banks, S., Armstrong, A., Booth, M., Brown, G., Carter, K., Clarkson, M., Corner, L., Genus, A., Gilroy, R., Henfrey, T., Hudson, K., Jenner, A., Moss, R., Roddy, D., and Russell, A., 2014. Using co-inquiry to study co-inquiry: community-university perspectives on research collaboration. *Journal of Community Engagement and Scholarship*, 7 (1), 37–47.

Berkes, F., 2009. Evolution of co-management: role of knowledge generation, bridging organizations and social learning. *Journal of Environmental Management*, 90 (5), 1692–1702.

Borda, O. F., 2001. Participatory (action) research in social theory: origins and challenges. *In*: P. Reason and H. Bradbury, eds. *Handbook of action research*, London, Thousand Oaks, CA and New Delhi: Sage, 27–32.

Braun, B., and Whatmore, S., 2011. *Political matter: technoscience, democracy and public life.* Minneapolis, MN: University of Minnesota Press.

Bresalier, M., Cassidy, A., and Woods, A., 2015. One health in history. *In*: J. Zinnstag, ed. *One health*. Croydon: CAB International, 1–15.

Burch, G.E., 1959. Of the normal dog. *The American Heart Journal*, 58 (6), 805–806.

Candea, M., 2013. Habituating meerkats and redescribing animal behaviour science. *Theory, Culture & Society*, 30 (7–8), 105–128.

Chatterton, P., and Pickerill, J., 2010. Everyday activism and transitions to postcapitalist worlds. *Transactions of the Institute of British Geographers*, 35 (4), 475–490.

Collard, R., and Gillespie, K., 2015, *Critical animal geographies*. London: Routledge.

Cornwell, J., 2011. Worker co-operatives and spaces of possibility: an investigation of subject space at collective copies. *Antipode*, 44 (3), 725–744.

Davies, G., 2012. Caring for the multiple and the multitude: assembling animal welfare and enabling ethical critique. *Environment and Planning D: Society and Space*, 30 (4), 623–638.

Degeling, C., 2008. Canines, consanguinity and one-medicine: all the qualitites of dog except loyalty. *Health and History*, 10 (2), 23–47.

Despret, V., 2004. The body we care for: figures of anthropo-zoo-genesis. *Body & Society*, 10 (2–3), 111–134.

Despret, V., 2013. Responding bodies and partial affinities in human-animal worlds. *Theory, Culture & Society*, 30 (7–8), 51–76.

Döring, D., Haberland, B.E., Ossig, A., Küchenhoff, H., Dobenecker, B., Hack, R., Schmidt, J., and Erhard, M.H., 2016. Behaviour of laboratory beagles: assessment in a standardised behaviour test using novel stimuli and situations. *Journal of veterinary behaviour*, 11, 18–25.

Dror, O., 1999. The affect of experiment: the turn to emotions in Anglo-American physiology, 1900–1940. *Isis*, 90 (2), 205–237.

Elston, M.A., 1987. Women and anti-vivisection in Victorian England, 1870–1900. *In*: N. Rupke, ed. *Vivisection in historical perspective*. New York: Routledge, 25–294.

EU Directive 2010/63/EU of 22 September 2010 on the protection of animals used for scientific purposes.

French, R., 1975. *Antivivisection and medical science in Victorian society*. Princeton, NJ and London: Princeton University Press.

Giraud, E.H.S., and Hollin, G.J.S., 2016. Care, laboratory beagles and affective utopia. *Theory, Culture & Society*, Epub, DOI: 10.1177/0263276415619685.

Greenhough, B., 2014. More than human geographies. *In* R. Lee, R., Castree, N., Kitchin, R., Lawson, V., Paasi, A., Philo, C., Radcliffe, S., Roberts, S.M., and Withers, C.W.J., eds. *Sage handbook of human geography*. London, Thousand Oaks, CA and New Delhi: Sage, 94–119.

Greenhough, B., and Roe, E., 2011. Ethics, space, and somatic sensibilities: comparing relationships between scientific researchers and their human and animal experimental subjects. *Environment and Planning D: Society and Space*, 29 (1), 47–66.

Haraway, D., 1997. *Modest_witness@second_millennium. FemaleMan©_meets_oncoMouse*™. New York and London: Routledge.

Haraway, D., 2008. *When species meet*. Minneapolis, MN: University of Minnesota Press.

Harbers, H., ed., 2005. *Inside the politics of technology*. Amsterdam: Amsterdam University Press.

Hinchliffe, S., Kearnes, M.B., Degen, M., and Whatmore, S., 2005. Urban wild things: a cosmopolitical experiment. *Environment and Planning D: Society and Space*, 23 (5), 643–658.

Jasanoff, S., 2004. *States of knowledge: the co-production of science and the social order*. London: Routledge.

Johnson, E., 2015. Of lobsters, laboratories and war: animal studies and the temporality of more-than-human encounters. *Environment and Planning D: Society and Space*, 33 (2), 296–313.

Joint Working Group on Refinement, 2004. Refining dog husbandry and care. Eighth report of BVAAWF/FRAME/RSPCA/UFAW joint working group on refinement. *Laboratory Animal*, 38 (Suppl. 1), 1e94.

Kirk, R., 2010. A brave new animal for a brave new world: the British laboratory animals bureau and the constitution of international standards of laboratory animal production and use, circa 1947–1968. *Isis*, 101 (1), 62–94.

Kirk, R., 2014. The invention of the 'stressed animal' and the development of a science of animal welfare, 1947–86. *In:* D. Cantor and E. Ramsden, eds. *Stress, shock, and adaptation in the twentieth century*. Rochester, NY: University of Rochester Press, 241–263.

Lederer, S., 1987. The controversy over animal experimentation in America, 1880–1914. *In:* N. Rupke, ed. *Vivisection in historical perspective*. London: Croom Helm, 236–258.

Lederer, S., 1992. Political animals: the shaping of biomedical research literature in twentieth-century America. *Isis*, 83 (1), 61–79.

Lee, N., 2008. Awake, asleep, adult, child: an a-humanist account of persons. *Body & Society*, 14 (4), 57–74.

Lorimer, J., 2007. Nonhuman charisma. *Environment and Planning D: Society and Space*, 25 (5), 911–932.

Lorimer, J., 2015. *Wildlife in the anthropocene*. Minneapolis, MN: University of Minnesota Press.

McKelvie, D.H., and Andersen, A.C., 1966. Production and care of laboratory Beagles. Journal of the Institute of Animal Technicians 17 (1), 25–33.

McKelvie, D.H., Andersen, A.C., Rosenblatt, L.S., and Bustad, L.K., 1971. The standardized dog as a laboratory animal. *In*: National Academy of Sciences, ed. *Defining the laboratory animal*. Washington, DC: National Academy of Sciences, 628.

Ostrom, E., 1990. *Governing the commons*. Cambridge and New York: Cambridge University Press.

Ostrom, E., 1996. Crossing the great divide: co-production, synergy and development. *World Development*, 24 (6), 1073–1087.

Pickerill, J., and Chatterton, P., 2006. Notes towards autonomous geographies: creation, resistance and self-management as survival tactics. *Progress in Human Geography*, 30 (6), 730–746.

Pickersgill, M., 2012. The co-production of science, ethics and emotion. *Science, Technology & Human Values*, 37 (6), 579–603.

Roe, E., and Greenhough, B., 2014. Experimental partnering: interpreting improvisory habits in the research field. *International Journal of Social Research Methodology*, 17 (1), 45–57.

School of Veterinary Medicine, U. of C., 1961. Tenth annual progress report. AEC Project No. 4.

Scott, J., 1970. A laboratory breed. *Science*, 170 (3959), 723.

Solarz, A.K., 1970. Behavior. *In:* A.C. Andersen, ed. *The beagle as an experimental dog*. Ames, IA: Iowa State University Press, 616.

Stengers, I., 2005. The cosmopolitical proposal. *In:* B. Latour and P. Weibel, eds. *Making things public: atmospheres of democracy*. Cambridge, MA: MIT Press, 994–1003.

Stengers, I., 2010. *Cosmopolitics I*. Minneapolis, MN: University of Minnesota Press.

Stengers, I., 2011. *Cosmopolitics II*. Minneapolis, MN: University of Minnesota Press.

Thompson, R. C., 1989. *Life-span effects of radiation in the beagle dog*. Report prepared for U.S. Dept. of Energy, Office of Health and Environmental Research, PNL-6822, UC-408.

Tomkins, L. M., Thomson, P. C., and McGreevy, P. D., 2011. Behavioral and physiological predictors of guide dog success. *Journal of Veterinary Behavior*, 6 (3), 178–187.

van Dooren, T., 2014. *Flight ways*. New York: Columbia University Press.

Willett, C., 2014. *Interspecies ethics*. New York: Columbia University Press.

Whatmore, S., 2006. Materialist returns: practising cultural geography in and for a more-than-human world. *Cultural Geographies*, 13 (4), 600–609.

Zinn, R. D., 1968. The research dog. *Journal of the American Veterinary Medical Association*, 153 (12), 1883–1886.

11 Rethinking ethnobotany? A methodological reflection on human-plant research

Jennifer Atchison and Lesley Head

Introduction

All human life is sustained by plants. Our bodies, our livelihoods, our futures are immersed in the thick oxygenated space between plants and the sun (Morton 2009). Plants underpin our food supply and contribute to the air we breathe, but this is not a simple relationship of dependency. Given the current predictions for global climate change for example, we might now argue that human–plant futures are mutually concerned in a way that has no historical precedent. Given this mutual entanglement of humans and plants, how might we trace the ways that plants also enter into research practices? In particular, and as this edited collection encourages us to ask, what might it mean to consider how we coproduce knowledge with plants? Since plants are central to pressing sustainability issues – biodiversity loss, climate change, and food security for example – how much should we centre them in our research endeavours on these issues? And how should we do so? Is it just an anthropomorphic conceit to consider that we can engage with them as other than objects, or that the research relationship might be one of mutuality and collaboration? Thinking through these questions at the boundaries of more-than-human and participatory research literatures has been challenging for us. It is one thing to recognise that the choice of methods can privilege some human voices over others, and think of ways to deal with this; it is quite another to try and give 'voice' (Plumwood 2009) to plants themselves. In this chapter, we discuss why a consideration of plants as active participants within the research process is inherently difficult, while also acknowledging that useful things emerge when we are pushed to think through the implications of this idea. Thus, while we are enormously sympathetic to exploring the methodological implications of more-than-humans as collaborators, we also want to bring a critical eye to these discussions.

In this chapter, we consider these questions by reflecting on aspects of our own research trajectory spanning two decades and three different Australian projects, each containing an ethnobotanical component: entanglements with yams in the east Kimberley, following wheat in southern New South Wales, and living with weeds across the tropical north. The human–plant entanglements in these examples encompass both mutual flourishing and extremely adversarial relationships.

We have no desire to expound a vegetal romanticism, much as we love trees. The brutal contingencies of continental plant invasions and biodiversity loss demand much broader and systematic thinking. At the same time, we are alert to the dangers of further asserting hierarchies of knowledge. As some of the most radical rethinking of plant capacities is coming from challenging and contested work in the botanical sciences, where more-than-human modes of sensing provide new insights into plant worlds, whether or not humans are part of them, we need to be alert to the dangers of asserting the power of science, but still be open to the radically new knowledges it is currently producing.

The backgrounding of plants: an example from ethnobotany and recent shifts in recognition

The first part of the challenge of considering plants as potential collaborators in research processes is to recognise how profoundly they have been backgrounded in Western thought (Hall 2011). Most of their contribution to our everyday lives goes unnoticed – plants are everywhere but nowhere. Since Aristotle defined animals as those who move and plants as those that do not, this plane of difference has become hierarchical, with animals assumed to be superior beings. The task of undoing such hierarchies and foregrounding the plants, animals, and people in the background, is therefore a call to action shared by participatory research methods and more-than-human research. However, in the long and diverse traditions that have studied human–plant relationships, the dominant research framings have been human oriented.

Ethnobotany provides a prime example as a field which has positioned itself as the science of the relationship between people and plants. Here too we find an orientation toward the human. Somewhat paradoxically, this is arguably strongest in studies of indigenous and non-Western modes of relating to plants. Even where indigenous people may interact with plants in a way that acknowledges their sentience, agency, and/or subjectivity, the dominant *researcher* interest has been in the implications for human sociality (Nolan and Turner 2011). Thus, while ethnobotany has contributed to opening up knowledge of the world beyond dominant Western world views, at the same time, it stands accused of privileging these same knowledge systems through methods which re-articulate the 'objects' of study.

Consider the rethink that ethnobotany has been undergoing over the past five years. A major demographic transformation of practitioners, with a predicted majority of non-Western graduates in U.S. natural and biological science university programmes by 2020, has prompted discussion there about the future of ethnobotany, including a more explicitly politically engaged trajectory. In what has been notionally called 'Ethnobotany 5' – Gary Nabhan and colleagues have articulated some of the key empirical and theoretical aspects that the field might encompass; a seamless gradient between professional science and citizen science, a predilection to do applied science in the service of environmental and social justice, and attention to novel ethical approaches (Nabhan *et al.* 2011b, p. 173). A recent compilation from South America (Albuquerque *et al.* 2014a) illustrates the different ways in which scholars have taken up 'Ethnobotany 5', including innovative approaches to thinking about theory and method in non-traditional spaces (Hurrell and Pochettino 2014), as well as ongoing epistemological issues in producing and practicing scientific research with communities (Albuquerque *et al.* 2014b). Notably, all these suggestions for rethinking the discipline focus on relationships between humans, leaving it unclear whether there is an active role for the botany in ethnobotany.

Arguably, the dominant drivers in ethnobotany's warranted rethink are serious human–human problems, including erosions of trust between indigenous people and researchers, misappropriation of knowledge, and exploitation by bioprospecting companies. Nabhan *et al.* have asserted that it is time to wed concern about contemporary ecological and social issues and insights from political ecology with ethnobiology/botany, a research field traditionally divorced from thinking about the 'global and macro-economic pressures on so-called traditional societies' (2011a, p. 1). The 'radical' trajectory envisaged here is a proudly human-centric one, explicitly aiming for sustained and critical engagement with people and their relationships with the more-than-human world, rather than with 'things' like plants in and of themselves (Nabhan et al. 2011a). The second part of the challenge of thinking of plants as potential collaborators, then, is to recognise that work to foreground plants is usually enmeshed in efforts to undo other power relations and established hierarchies, such as those between different groups of people. Recognising and elaborating whatever concerns plants may have, is not always an obvious or pressing task for research.

The relevance of participatory research?

Here we focus on the production of knowledge and whether, and in what ways, we might centre plants in our research process. We situate our anxieties in the overlapping discussions of plants coproducing research and the explicit and direct consideration of plants as co-creators, or research collaborators.

The concept of coproduction, attributed to Ostrom (1996), in reference to shifting attention to the active agency of those who might be dismissed as passive, has gained new resonance in science and technology and also more-than-human studies, in particular as a critique of 'realist ideology' (Jasanoff 2004, p. 3). In the contested interactions between people, ideas, institutions, and material objects, Jasanoff (2004) has argued that the idiom of coproduction provides explanatory power, enabling us to think about the natural and the social as being produced together (Latour 1993). In particular, coproduction offers new ways of thinking about power by highlighting the often invisible 'knowledges, technical practices and material objects in shaping, sustaining, subverting or transforming relations of authority' (Jasanoff 2004, p. 4). It does this in two ways, first by demonstrating how power frames and organises knowledge but is simultaneously expressed through it, and second by insisting that 'power is constituted as much through elision of marginalised alternatives as through the positive adoption of dominant viewpoints' (Jasanoff 2004, p. 280). The 'strategic silences' thus become the research interest or grist for the mill.

Participatory research methods on the other hand, place emphasis on shifting from research subject to co-researchers. Here, participants contribute to the building and testing of research concerns in collaboration with academic researchers, as they seek to 'democratise knowledge production and foster opportunities for those involved' (Kindon *et al.* 2008, p. 90). Collaboration is an 'explicit' orientation throughout the research practice. Researchers within anthropology, for example, have focused on collaboration for a range of reasons, including to 'engender texts that are more readable, relevant and applicable to local communities' (Lassiter 2008, p. 73). The insistence in going beyond involvement and/or engagement to focusing on active participation, has a long history, and has been imperative in highlighting both the political right of people to be involved in research as well as their having a say in processes and decision making which may affect them (Borda 2001).

While never straightforward and certainly unevenly applied, both coproduction and participatory research have in common the emphasis on participation in research as an 'iterative process of overcoming power imbalances' (Beebeejaun *et al.* 2014, p. 40). For us, the difference and difficulties lie within the emphasis on the 'team' or 'joint endeavour' within participatory research processes. That is, to what extent we might go beyond acknowledging that plants have agency and affect the research process, and that thus research is or can in Jasanoff's sense be coproduced with them, to considering plants as collaborators with an interest in, and capacity to act as co-creators of research as practitioners of participatory methods understand it to be. Our discomfort here lies with the implicit frame of a human norm in the research process, and our reluctance or inability to say that we know enough of plants to be adequately inclusive. But there are additional reasons why making plants central to our research process is not straightforward.

First, perhaps counter-intuitively, the planes of difference between humans and plants are not self-evident. Elsewhere, we have detailed how human geographers, and others, have contested the idea and practice of human exceptionalism and are rethinking human and nonhuman identity and subjectivity, along with attempts to understand the differences within networks of particular beings/things (Head *et al.* 2014, p. 2, Head *et al.* 2015). The implications of these differences are yet to be worked through, to the extent that they have been, for example, in animal studies. Second, and relatedly, plant differences, like those of other nonhumans, are not only mediated through their relationships with humans (Smart 2014, p. 4). Building on a rich history of feminist approaches to difference and to specificity, we have argued (like Jones and Cloke [2002] and Lulka [2009] before us), that it is necessary to build depth to the nonhuman, including as those lives are lived independently from us. This parallels efforts in animal studies to consider lives as lived away from the human gaze, in order that these lives might be made more visible (Bear 2011). Here we reiterate that nonhuman capacities, including for participation within research, must be allowed to emerge and be examined in empirical specifics (Head et al. 2015).

Third, key elements of research would require innovation, including the languages used, research concerns and methodological approaches. For example, while researchers have pursued different methodologies in their attempts to be more attentive to nonhumans, such as story-telling, walking (Sundberg 2014) or performance, visual or non-textual approaches (Beebeejaun et al. 2014), and being attuned and oriented to plants presents particular methodological challenges. On the one hand, 'bringing the animals (and plants and microbes and tools) back in' reflects a rich heritage within anthropology of study of humans and more-than-humans through focusing and attending to everyday and embodied interactions between people and other things (Smart 2014, p. 4). These methodologies remain important but may need to be tailored to plants (Pitt 2015). On the other hand, considering the distinctive and independent capacities of plants (e.g. the treeness of trees: Jones and Cloke 2002), requires innovation and botanical sensing which extends our own human capacities.

In the following section, we reflect on these methodological challenges by briefly recounting three encounters with particular plants: yams, wheat, and invasive rubber vine, each a focal plant in three large Australian research projects over the past 25 years. Although presented chronologically, these projects and the plants we encountered continue to reverberate in our thinking and writing. These reflections highlight how our thinking has changed, as well as which methodologies have helped orient us as researchers to different human-plant relations.

Encountering plants

In the yam hole

In the cool dry winter months of northern Australia, Aboriginal women and children look for edible yams in the small isolated rainforest patches across the savannah. Women search out the, by now, dried up, often fragmented, remnants of vines amongst the tangle of trees and rocks and differentiate them from other superficially similar but unrelated climbers. To do this, a yam digger relies on her knowledge of the shape and colour of stems; Mamunya (*Dioscorea bulbifera*) stems are sometimes slightly ridged, Kagaoli (*D. transversa*) stems often striated. These stems provide the important physical connection from the aerial part of the plant to the subterranean tuber and it is vital that the digger persist in maintaining and following this connection through the digging process. If that connection is broken or lost, the digger must search for it or begin again.

Our lessons in yam collecting began 25 years ago in the Keep River region of the eastern Kimberley, Northern Territory. We accompanied senior Aboriginal women, younger women, and children on dozens of regular trips to collect yams – as either the main activity of the day or as part of other hunting or fishing activities. Digging yams in this part of the country is a labour-intensive process. In rocky country, women must move large volumes of dirt and rock and cut away thick tree roots. Women might spend hours looking and digging only to retrieve very little or even nothing. Over repeated days of digging for yams, it was evident that this was also a time in which the women might sit quietly in the shade and think or talk.

Yam plants, the soil substrate and the rocks amongst which they grow are folded together into particular relations with the people who collect them. There is an emergent attentive watchfulness and quietening of those digging and watching as the yam gradually reveals itself. Other processes of daily life go on and are also brought into being through the process of collecting yams. Collecting yams requires people to be patient. These skills of attentiveness, patience, quietening, and persistent digging must be learnt over time, and carried out within the processes of daily life, across yearly seasonal changes and even over longer cycles of renewal and regeneration.

To dig for yams and to be sustained by them is to be physically implicated; to feel between fingers and thumb for the ridges or striations, to scrape away the dirt and rocks and to persist. Each yam, each stem, each hole is different but over time a digger becomes attuned to the yam and its habits, to yam habitats and their people. Regular collecting activity by Aboriginal people seems to be implicated in the ongoing maintenance of quality yams over the longer term (Head et al. 2002). The top of a freshly dug yam is placed in the bottom of the hole over which leaf litter, loose soil, and any water runoff collects and sustains the new vine shoots that grow from the piece of yam over the following year.

The yam hole is a space in which yams and the people who search for them are bound, where yams and people become together. The hole is of course, empty, but it imbricates both what has gone before and what is yet to come. For our Aboriginal participants, to see old yam holes is to know of people and yams past, to see and recognise what has been before and that which is to yet come. As researchers we were eventually expected to find our own yams. Our position as observers was tolerated for only a limited time; since to be present is to be called into action – to sit and dig and create a yam hole for ourselves.

On reflection, it is fair to say that our vigilance, attentiveness, and listening in this project was directed much more at our human relationships than to the plants themselves. Nervously entering the fraught world of indigenous research collaborations, we were constantly on guard to develop and maintain appropriate ethical relations with the people.

In our case, we did not direct comparable levels of concern to the plants; they were scientific samples to be observed and collected, identified, and examined back in the laboratory. Rather, we came to this research with biogeographic and archaeological training, and scientific field skills that emphasised systematic sampling of various kinds. Our training taught us to sample plants in specific material ways – pick them, press them, label them, name them, and assemble them into different kinds of data, as well as to consult the botanical sensing of others who had come to know them (Atchison and Head 2013a). The intellectual context in which we were writing was one in which it was still necessary to assert the agency of human activities – as opposed to the forces of 'nature' – at a landscape scale (Head et al. 2002, Atchison 2009).

At the supermarket

Some years later, in a different project, we wondered how to approach ethnobotany among people whose relations with a plant (wheat) is via the supermarket and a picture on the cereal packet before breakfast (Head et al. 2012). Influenced by the work of Ian Cook et al. (2006), and the methodology we have come to know as 'following', we found ourselves following wheat into all sorts of nooks and crannies. In this study, the question of coproduction was not at issue; domesticates are widely understood as our closest plant relations. Nevertheless, the methods were challenging, partly because of the way the plants changed form – what exactly were we to follow?

The contemporary industrial transformation of wheat takes it beyond its very visible status as a staple food, important as that is (Atchison et al. 2010). It can be pulled apart and re-assembled in many different ways. If you were reading your breakfast cereal box this morning – you may have been aware of the fact that you were eating a plant. However it is unlikely that the packaging would have drawn attention to this fact if you were feeding your cat or washing your hair – since to do so would render the industrial nature of their production visible. The many transformations of wheat challenge us as researchers to be attuned to the invisible status of plants – even to those that are familiar and everyday.

In the supermarket, the food label is a key research tool as well as a significant agent in the assemblage that connects contemporary city dwellers to the conditions under which their food is produced. With varying degrees of care, most people will read a food label of regularly purchased products – the ingredient list, country of origin, and nutritional information are all embedded here (Atchison et al. 2010). For sufferers of coeliac disease however, who must be highly attuned to its presence and absence because of the potential danger it represents, the food label is vital. In Australia, current labelling laws require wheat to be identified in a food ingredient list – for example starch, or food additive, or thickeners derived from starch must all be declared as food items containing wheat derived products.

With the help of food ingredient labels designed to assist coeliac sufferers, we mapped and traced wheat across the supermarket. Wheat was very visible in food products with 29.8% of all food items containing wheat, but very likely present, but rarely declared (4%), in many non-food products. In fact, wheat is a very common ingredient in a range of non-food items, as our interviews with industrial chemists revealed. Wheat flour is transformed in the starch factory into washed gluten, modified starches and fibre. These end up in paper, cardboard, shampoo, nutritional supplements, cosmetics, and pet food, amongst other things. In pet food for example, wheat is not just an ingredient, it is vital to the extrusion process which gives dry pelletised dog biscuits their shape. In the case of milk, where wheat constitutes up to 50% of the feed for cows in some Australian dairy herds, it was not even an ingredient, but instead becomes the energy feeding the system, fuelling and reenergising as it goes (Head et al. 2012).

The supermarket unfurls the industrial material transformation of plants and their connections to everyday life. As a research site, its shifting assembly and re-organisation of food stuffs and other items make some wheat visible in certain forms, but not others (Atchison et al. 2010). The supermarket collects, assembles, and disperses these materials, but it does so together with food packaging, which itself makes visible particular wheat histories, spatialities, and trajectories. Food packaging also tells us something of the ways in which plant identities become fixed and unfixed. Although industrial transformations make it harder for consumers to recognise and see the ingredients or processes behind their purchases – they do not disconnect us from wheat. In fact wheat now transforms so many aspects of our lives we might say that we are more connected to it than ever before.

Like tenacious shoppers, we as researchers had to learn new skills in the supermarket, including understanding the layout of isles and shelves, but also reading food labelling, relearning organic chemistry, and pursuing wheat 'experts' into different arenas to attune to this industrial transformation. Our 'attuning' to plants in this case was an active and interrogative, mobile engagement mirroring the energetic power of the wheat itself. In this project, we were alert to the role of science in providing insights into plant worlds. The industrial chemists for example, understood that not all wheat was the same, but had different qualities and constituent components of similar, but not identical, chemical assemblages; gluten, starch, polysaccharides, which were themselves assemblages of other organic molecules. These were broken apart in particular ways, concentrated and then isolated for use elsewhere. Thus, understanding the chemical scientists' disassembly of wheat enabled us to trace gluten into food and places where the wheat would otherwise have been invisible, in the shampoo and fabric conditioner, but also to recognise its reassembly as the new 'shape' in the dog biscuit, or 'mouth-feel' in the ice-cream.

These scientific insights into plant worlds were gleaned from chemical engineers who have come to know wheat in laboratories, under microscopes, and through the industrial machinery of starch factories and industrial food chemistry. As 'expert' as these scientists are, we recognise that that such insight is brought into being with plants in specific ways that are laden with power and privilege over others. Such ways of scientifically knowing the world are dominantly constructed and construed as authoritative. We recognise there is a real danger here, that in highlighting the capacity of science to expand ways of knowing the world, we are potentially complicit in re-asserting that power hegemony.

This project influenced our own trajectory of thinking about the role of science and our capacity as scientists to generate new insights into plant worlds. As bio-geographical scientists, we maintain that scientific ways of knowing plants generates particular and useful insights, but that these knowledges and insights are different, not necessarily better, than other ways in which plants are understood. It is also important to note here that plant science is itself in a state of flux, with some plant scientists critiquing the disciplinary hegemony of others. Work from biology on plant learning and sensing by Gagliano *et al.* (2014), for example, requires us to go beyond simplistic questions of whether or not plants have agency, to a more complicated recognition of agency by thinking about the ways in which distinct nonhuman capacities are differently enacted (Head et al.

2015). Gagliano et al. (2012) recently illustrated previously uncharted signalling and sensing pathways by which plants communicate with each other, prompting calls for a newly integrated field of plant behavioural ecology and cognitive biology. That concepts like plant communication or plant intelligence (Firn 2004) are controversial within scientific realms continues to remind us that the terminology and reference points dominant in Western research cultures are still very much centred around the Aristotelian hierarchy between humans, animals, and plants.

In the eradication and control zones

Invasive alien species, as they are named in the adversarial style of the dominant management paradigm, are seen as a significant threat to global biodiversity and the viability of agriculture and other human enterprises (McGeoch et al. 2010). In this final example, we draw on our ethnographic project of invasive plant management in northern Australia to think about the specificity of plants and our trajectory of thinking through human–plant relations. Eradication zones demarcate highly significant spaces where coordinated efforts take place to contain prioritised plants. Control zones, on the other hand, denote more specific aims such as asset protection or improvement of pastoral production values. These zones are often conceptualised in invasive species management at the macro-scale of landscape, using maps of frontiers and continental discourse of war. Instead, we have shown that work to eradicate or control plants is undertaken at the micro or bodily scale, where human, animal, and plant bodies interact (Atchison and Head 2013b).

In the eradication zone in north western Australia, we talked to helicopter pilots who have learnt to spot and distinguish individual Rubber vine (*Cryptostegia grandiflora*) plants from the air by the reflection and colour of new growing leaves at a particular time in the early wet season. The plants growth response to rain is recognised by the helicopter pilots, who fly at low altitude, with the sun behind them, in order to pick up the reflection of leaf gloss and distinguish newly growing leaves from the shrubs over which they grow. In contrast, in the control zone in western Queensland, we flew with other pilots as they carried out incendiary bombing of contiguous Rubber vine amassed over hundreds of square kilometres, using ignited gelled petroleum slung below a helicopter. We watched here as pilots flew in concentric arcs around the vine towers, fanning the ignition fuel with the chopper blades, heating the fire up to over 600 degrees Celsius, at which point the latex in the vine would ignite and kill the plant. We focused on the material interactions between humans and rubber vine in these different settings as land managers set about trying to locate, kill, or control. In doing so, we used the conceptual device of bodies to think about what people were doing and the distinctive capacities of plants in each circumstance.

Most writing about bodies tends to assume that the bodies under consideration are human, or at least animal (Longhurst 2011). Plant bodies take many different forms; grass, tree, shrub, vine. Some plants (angiosperms) alternate morphological forms between adult and seed in their life-times. On the one hand, the differences between plant and other bodies are not intrinsic, but relational – stabilising and changing into particular forms. On the other hand, plants share a distinctive and particular assemblage of capacities that have come into being over evolutionary timescales. These distinctive capacities include the ability to photosynthesis, for example, or to form and materialise cellulose within cellular walls. We have argued that we need to be attentive to these distinctive capacities in order to rethink relationships with them, including concepts often framed against a human or animal norm (Head *et al.* 2015).

By focusing on the bodily scale in invasive plant management, we have shown that much land management effort to eradicate or control weedy plants takes place in the intimate spaces between plant and human bodies; between land managers who are compelled to kill them and the plants, who, by virtue of their complex life histories, mobile trajectories, and capacity to alternate, seemingly defy attempts to control them. Despite the clear rhetoric of eradication or control, weed management zones are spaces whose outcomes are always uncertain. Any sense of certainty about killing a single rubber vine plant, for example, is constantly overridden by the uncertainty about whether there will be more plants; more seeds in the seed bank capable of germinating, or more seeds blowing in on the wind. Once a plant has more obviously become a collective, grown, flowered, and reproduced itself, eradication is no longer a possibility, since eradication implies complete removal or destruction.

Focusing on plant bodies exposes some of the tensions and contradictions of Western weed management frameworks. It also exposes the different temporalities that are inflected through plants; the long duration of seed dormancy or the punctuated growth of opportunistic sprouting. These different temporalities in combination with other actors – tourist bodies or stock for example, together provide a different view of plant agency usually focused upon in the landscape scale. Seeds travelling in stock or animal hair are different to dispersal by annual floodwaters across riparian habitats. Together, these specific capacities, as they are inflected relationally, might give us different tools to monitor and manage the spread of invasive species, or at least help us think though the influencing variables.

The spaces of invasive species management exposes bodily encounters with humans and plants, but they also highlight the point that struggles for living and dying are mutual concerns, humans and plants alike. Invasive plants in particular have extended our own thinking about the agency and mobility of plants, but they have also stimulated us to think about their radical alterity, which challenges our very human, and even our animal-centric, notions of what a body is and what it can do. Individual and collective subjectivities are called into question. In this way, plants challenge our current configurations of the self and subject, the way in which bodies resist or become mobile, or act in the world.

Conclusions

Wright et al. (2012) have argued that taking relational ontology seriously requires us to be both attentive to and aware of the ways in which we are coproduced. There are multiple urgent challenges that will require us to consider this mutuality and our relationship with plants, in particular, biodiversity loss and climate change, amongst others. And, like the uncertainty embodied by the plants themselves, we will need to consider, perhaps more humbly, the diversity and capacities of these nonhumans with whom we cohabit the earth.

The implications for ethnobotany are no less uncertain. While most of us are used to considering the ethical implications of our work with human participants, and to a lesser extent research work that involves animals, thinking about how research might be collaboratively produced with plants is for us a much more difficult prospect. Recent work by Sarah Wright and colleagues, with Bawaka Country as author, is one example of the way in which researchers are grappling with human and nonhuman subjectivities and participation within research (Wright et al. 2012, Bawaka Country et al. 2015). This builds on previous scholarship, for example, as elaborated by Rose (1996) in the subjectivity accorded Country. In Bawaka Country, methodological concerns are ontological, shaping the politics and ethics of the research. For Wright *et al.*,

> Research encounters occur in sites of human and nonhuman coexistence and overlapping territories. Therefore, human and nonhuman beings can provide shape to research, making activities possible or impossible, prompting certain topics of discussion and closing others, generating shared moments and highlighting differences.
>
> (2012, p. 51)

Bawaka et al. (2015) provoke us as knowledge producers to consider research as already and always coproduced, an effort to destabilise both academic power and human authority. But we, as authors, are challenged here to go beyond illustrating the agency of country, or in our case plants, to discerning the specifics. Would it be possible to sense this coproduction in ways unmediated by people? How would the process of research itself change? Part of being situated researchers – and attending non-hierarchically to radical difference – is to recognise the limits of our own human sensing capacities; there are limits to what can be deduced from sitting beside a tree.

How has our thinking changed over time, and why? To capture some of the continuities – working with Aboriginal women has been particularly formative. It has pushed us to think more inclusively, to engage with the country, and to think about the assemblage of which we are but a part. Arguably, this led us to a better recognition of human perceptions of more-than-human agency, rather than a better recognition of yams directly, although thinking and writing through obligations to our human participants highlights this as an ongoing experience.

The methodology of 'following' and the archaeobotanical method to be attuned to absences have also been influential, the traces and windows offer us new openings to places such as 'the everyday' which we have tried to approach with fresh eyes. Bringing both of these methods together is reliant upon botanical scientific modes of sensing, dissecting, and organising. This includes scientific methods of observation and detachment, as well as active mobile methods of participating and attuning. In a sense then, we need to assert the capacity of science to expand human worlds/sensing, without suggesting these are the only or even the authoritative ways of knowing the world.

For embodied geographies, planty perspectives open up new ways of thinking about the multiple ways in which humans and plants are entangled – across landscapes, in daily life, and between bodies and their boundaries. Plants challenge us as researchers by expressing different forms of collectivity, mobility, and agency and (in the case of invasive plants) by suggesting very different frameworks to think about a future of having to 'live with' them. Recognising our mutual concerns will depend upon our capacity to illuminate differences as well as our interdependencies, but this is methodologically challenging. If 'the practices of knowing cannot be fully claimed as human practices' (Barad 2003, p. 829), there is still much to consider in the way that our research encounters with plants shape our thinking, and the research process.

References

Albuquerque, U., Cruz da Cunha, L., de Lucena, R., and Alves, R., eds., 2014a. *Methods and techniques in ethnobiology and ethnoecology*. New York: Humana Press.

Albuquerque, U., de Sousa Araujo, T., Soldati, G., and Fernandes, L., 2014b. 'Returning' ethnobiological research to the communities. *In:* U. Albuquerque, L. Cruz da Cunha, R. de Lucena, and R. Alves, eds. *Methods and techniques in ethnobiology and ethnoecology*. New York: Humana Press, 451–463.

Atchison, J., 2009. Human impacts on *Persoonia falcata*: perspectives on post-contact vegetation change in the keep river region Australia, from contemporary vegetation surveys. *Vegetation History and Archaeobotany*, 18 (2), 147–157.

Atchison, J., and Head, L., 2013a. Exploring human-plant entanglements: the case of Australian *Dioscorea* Yams. *In*: D. Frankel, J. Webb, and S. Lawrence, eds. *Archaeology in environment and technology: intersections and transformations*. New York: Routledge, 167–180.

Atchison, J., and Head, L., 2013b. Eradicating bodies in invasive plant management. *Environment and Planning D*, 31 (6), 951–968.

Atchison, J., Head, L., and Gates, A., 2010. Wheat as food, wheat as industrial substance: comparative geographies of transformation and mobility. *Geoforum*, 41 (2), 236–246.

Barad, K., 2003. Posthumanist performativity: toward an understanding of how matter comes to matter. *Signs: A Journal of Women in Culture and Society*, 28 (3), 801–831.

Bawaka Country, Wright, S., Suchet-Pearson, S., Lloyd, K., Burarrwanga, L., Ganambarr, R., Ganambarr-Stubbs, M., Ganambarr, B., and Maymuru, D., 2015. Working with and learning from Country: decentring human authority. *Cultural Geographies*, 22 (2), 269–283.

Bear, C., 2011. Being Angelica? Exploring individual animal geographies. *Area*, 43 (3), 297–304.

Beebeejaun, Y., Durose, C., Rees, J., Richardson, J., and Richardson, L., 2014. Beyond text: exploring ethos and method in co-producing research with communities. *Community Development Journal*, 49 (1), 37–53.

Borda, O.F., 2001. Participatory (action) research in social theory: origins and challenges. *In*: P. Reason and H. Bradbury, eds. *Handbook of action research participative inquiry and practice*, London: Sage, 27–38.

Cook, I., *et al.*, 2006. Geographies of food: following. *Progress in Human Geography*, 30 (5), 655–666.

Firn, R., 2004. Plant intelligence: an alternative point of view. *Annals of Botany*, 93 (4) 345–351.

Gagliano, M., Renton, M., Duvdevani, N., Timmins, M., and Mancuso, S., 2012. Out of sight but not out of mind: alternative means of communication in plants. *PLOS One*, 7 (5), e37382.

Gagliano, M., Renton, M., Depczynski, M., and Mancuso, S., 2014. Experience teaches plants to learn faster and forget slower in environments where it matters. *Oecologia*, 175 (1), 3–72.

Hall, M., 2011. *Plants as persons: a philosophical botany*. Albany, NY: SUNY Press.

Head, L., Atchison, J., and Fullagar, R., 2002. Country and garden: ethnobotany, archaeobotany and Aboriginal landscapes near the keep river, northwestern Australia. *Journal of Social Archaeology*, 2 (2), 173–196.

Head, L., Atchison, J., and Gates, A., 2012. *Ingrained: a human-biogeography of wheat*. Burlington: Ashgate.

Head, L., Atchison, J., and Phillips, C., 2015. The distinctive capacities of plants: insights from an adversarial relationship. *Transactions of the Institute of British Geographers*, 40 (3), 399–413.

Head, L., Atchison, J., Phillips, C., and Buckingham, K., 2014. Vegetal politics: belonging, practices and places. *Social and Cultural Geography*, 15 (8), 861–870.

Hurrell, J., and Pochettino, M., 2014. Urban ethnobotany: theoretical and methodological contributions. *In*: U. Albuquerque, L. Cruz da Cunha, R. de Lucena, and R. Alves, eds. *Methods and techniques in ethnobiology and ethnoecology*. New York: Humana Press, 293–310.

Jasanoff, S., 2004. The idiom of co-production. *In*: S. Jasanoff, ed. *States of knowledge: the co-production of science and social order*. London: Routledge, 1–12.

Jones, O., and Cloke, P., 2002. *Tree cultures: the place of trees and trees in their place*. Oxford: Berg.

Kindon, S., Pain, R., and Kesby, M., 2008. Participatory action research. *In*: R. Kitchin and N. Thrift, eds. *International encyclopedia of human geography*. Amsterdam: Elsevier, 90–95.

Lassiter, L., 2008. Moving past public anthropology and doing collaborative research. *NAPA Bulletin*, 29 (1), 70–86.

Latour, B., 1993. *We have never been modern*. Cambridge, MA: Harvard University Press.

Longhurst, R., 2011. *Bodies: exploring fluid boundaries*. London: Routledge.

Lulka, D., 2009. The residual humanism of hybridity: retaining a sense of the earth. *Transactions of the Institute of British Geographers* NS, 34 (3), 378–393.

McGeoch, M.A., Butchart, S.H.M., Spear, D., Marais, E., Kleynhans, E.J., Symes, A, Chanson, J., and Hoffmann, M., 2010. Global indicators of biological invasion: species numbers, biodiversity impact and policy responses. *Diversity and Distributions*, 16, 95–108.

Morton, O., 2009. *Eating the sun: how plants power the planet*. New York: HarperCollins.

Nabhan, G., Chambers, K., Tecklin, D., Perramond, E., and Sheridan, T., 2011a. Ethnobiology for a diverse world defining new disciplinary trajectories: mixing political ecology with ethnobiology. *Journal of Ethnobiology*, 31 (1), 1–3.

Nabhan, G., Wyndham, F., and Lep, D., 2011b. Ethnobiology for a diverse world: ethnobiology emerging from a time of crisis. *Journal of Ethnobiology*, 31 (2), 172–175.

Nolan, J., and Turner, N., 2011. Ethnobotany: the study of people-plant relationships. *In*: E.N. Anderson, D. Pearsall, E. Hunn, and N. Turner, eds. *Ethnobiology*. Hoboken, NJ: Wiley-Blackwell, 133–147.

Ostrom, E., 1996. Crossing the great divide: coproduction, synergy, and development. *World Development*, 24 (6), 1073–1087.

Pitt, H., 2015. On showing and being shown plants – a guide to methods for more-than-human geography. *Area*, 47 (1), 48–55.

Plumwood, V., 2009. Nature in the active voice. *Australian Humanities Review*, 46 (May), 113–129.

Rose, D., 1996. *Nourishing terrains: Australian aboriginal views of landscape and wilderness.* Canberra: Australian Heritage Commission.

Smart, A., 2014. Critical perspectives on multispecies ethnography. *Critique of Anthropology*, 34 (1), 3–7.

Sundberg, J., 2014. Decolonizing posthumanist geographies. *Cultural Geographies* 21 (1), 33–47.

Wright, S., Lloyd, K., Suchet-Pearson, S., Burarrwanga, L., Tofa, M., and Country, B., 2012. Telling stories in, through and with country: engaging with Indigenous and more-than-human methodologies at Bawaka, NE Australia. *Journal of Cultural Geography*, 29 (1), 39–60.

12 Con-versing

Listening, speaking, turning

Deirdre Heddon

Click

Introduction

> Everyone sits deep in thought. It is very quiet. All that can be heard is FIRS's low muttering. Suddenly a distant sound is heard. It seems to come from the sky and is the sound of a breaking string. It dies away sadly.
>
> (Chekhov 1904, Act Two)

> There must be some problem of listening if we only hear from the earth when it is so seriously endangered that we cannot help paying heed.
>
> (Fiumara 1990, p.6)

The focus of this chapter is a failed conversational moment staged in a clearing in a forest. This conversation took place as part of the research project, *In conversation with. . .: co-designing with more than human communities* (see Bastian 2013). That the purpose of the overall project was to explore methods for conversing with nonhuman actors (animals, insects, plants and elements) renders the dramatic failure of human-to-human conversation all the more ironic.

If participatory research is understood as a methodology that involves research partners fully in the knowledge-production process, then communication is clearly key to its success. Turning this around, methods of participatory research could be seen as useful to developing better conversations. Durham University's *Participatory Action Research Toolkit* (Pain et al. 2010–2011), set as a key reading for our research project, summarises the typical stages of participatory action research (PAR): Planning, Action, Reflection and Evaluation.[1] Each of these stages requires communication. Significantly, the words used most frequently across the different stages outlined in the toolkit are 'collaborative' and 'collective'. 'Participation' translated as 'collaboration' makes more transparent its articulation with notions of ethics, politics and power. My interest here is in exploring the relationship between collaboration, and conversational and communicative practices – both human and nonhuman.

Though the conversation between these humans in the forest was a failure, it has proved generative as a site for reflection. As our human attention turned to nonhuman co-researchers, we neglected to attend to the still-participating human co-researchers. In turning away from the human, we also turned away from human participatory methods, including the creation at the outset of agreed processes and safe spaces, open and non-judgemental attitudes, and horizontal and democratic structures of communication. Holding these strategies in mind, but extending them creatively, might have allowed for more transformative conversations with nonhumans too. My return to the scene of failure allows for a reflective exploration of the relationship of listening to speaking, and from there, the relationship of listening otherwise to speaking otherwise and to speaking other worlds. The argument I offer here is that different modes of listening afford different ways of speaking, thus opening conversations to, between and with the human and more-than-human.

Since the practices of listening are pivotal to my discussion, I adopt a performative listening device throughout this essay. As discussed below, in one of the workshops of *In conversation with. . .*, I tentatively forged my journey across a room by listening out for the sound of a click. The click – waiting for it, hearing it, not hearing it – kept me focused and attentive. I borrow this 'click' here in order to provide the reader with a similar mechanism for focusing and attending. The *click* is intended as an intervention that calls to attention; a reminder to listen without presuming to already know the direction or the move in advance. To already know is to kill curiosity. More significantly, it is to foreclose future worlds.[2]

Setting the scene

It is April 2013 and I am participating in the first workshop of the project. Funded by the Arts and Humanities Research Council, as part of their *Connected Communities* theme, the project's intention is to challenge the assumption implicit in the theme's funding call that 'communities' are singularly human formations and that co-researchers are therefore also (only) human.[3] As our project blog notes, both community and research 'have always been entangled with the lives, qualities and capacities of nonhuman actors' (Bastian 2013). Our ambition for the project is to bring nonhuman actors into the process of research as active participants, believing that this might, in turn, extend understanding of the methods by which research can be co-designed and co-produced. The human co-researchers in *In conversation with. . .* are drawn from across a range of arts, humanities and social science disciplines, including philosophy, cultural geography, sociology and creative arts. Other human co-researchers are artists, beekeepers and community outreach practitioners.[4]

In this first workshop in a series of four, the co-designers we human researchers are in conversation with are dogs – namely, specifically, Winnie and Cosmo, workers with Dogs for the Disabled.[5] They are supported by two other workers from Dogs for the Disabled, trainers Helen McCain and Duncan Edwards. McCain's companion dog, Willoughby, also participates in part of the workshop. The workshop has been designed by Michelle Bastian, a philosopher from the University of Edinburgh, and Clara Mancini, a researcher in Human/Animal–Computer Interaction at the Open University.

Sited in a seminar room on the Open University's campus at Milton Keynes (UK), the workshop draws on areas under development at the University's Centre for Research in Computing, particularly its research into Animal–Computer Interaction (ACI). ACI extends to animals the Human–Computer Interaction's focus on satisfying computer users' practical and subjective experiences – in simple terms: What do I need to do, how can I do it, and what does that feel like?[6] In our workshop, we are concerned to think about dogs specifically as I.T. users, since, as is made apparent during the workshop itself, dogs have already proved to be accomplished in completing tasks typically assumed to be for humans and by humans, including using technologies such as ATM machines.

The brief we are set by Bastian and Mancini at the start of the workshop is to think about how we might work *with* dogs on a design function. Spending most of the first day in the company of Winnie and Cosmo, observing them, interacting with them and attempting to 'be' them even, we decide that our focus on the second day will be to design either a door or a button which addresses dogs as end-users.

As I explain in the post-workshop blog:

> Here is the design problem: 'Design a button for dogs to use'. Let's translate that as 'design a better button', better for dogs and more-than-dogs. Working with dogs [during the workshop] has attuned us to the range and prevalence of buttons/switches in our everyday – the button that releases the door, the buttons on our phones, our remotes, our tablets, the switches on the sockets. We inhabit a world of buttons.
>
> We focus on the buttons that release doors. We reflect on how the dogs – Winnie and Cosmo – had to press their paw to the centre of the mock button before being rewarded (with clicker, treat and positive affirmation 'Good Boy!'). And how difficult this task is because it demands precision (paws hit the edge, slip off, miss the centre). And this leads us to recognise that the button design unnecessarily focuses attention on a hot spot. ([Co-researcher] Owain has recently downloaded an app to his phone – the Big Red Button app – which demonstrates the button-centre concept perfectly.) The button is a bull's eye, a dead centre, the apocalypse, lift off, jackpot. The button is a cultural repository. But it's also an unnecessary restriction and challenge.
>
> So, imagine the button is replaced by a long, vertical strip. There is no dead centre, no requirement for precision, for a steady and accurate finger or paw. This strip, an extended contact zone, extends the potential for connection between one thing and another.
>
> This is our proposed new design but we decide to put functionality aside for now and focus on pleasure. What would a dog like from a strip such as this (textures? smells? activities? challenges?)? What might make the task of contact one of pleasure, engagement, stimulation – beyond the reward of achievement itself (*Good Boy!*)? We will start with the dogs' experiences, rather than being led by the necessity of functionality, flipping the instrumental and the aesthetic. And for this, we will need to engage in some deep hanging about with dogs.
>
> (Heddon 2013)

As is acknowledged in this post, our arrival at the idea of the strip over the button – indeed, our consciousness even of 'the button' *as* a pervasive design feature in our 21st-century culture – is a direct result of the dogs' participation in the workshop. During the first day, in the company of Winnie and Cosmo, we learnt about what dogs who work for Dogs for the Disabled do. We also observed the form of training that precedes and enables this. The training used is the 'clicker method', whereby a click signals 'good', and in the training stage is always followed closely with an edible treat and vocal praise ('*Good Boy!*'). Click, treat and praise are intended to reinforce the desired behaviour – which might be to press a button to open a door, or pick up a remote control from the floor. To receive the praise/treat, the dog must listen out for, and respond to, the clicker. Putting aside any ethical consideration of this method of conditioning, I am not the only one to profess admiration in the face of Winnie and Cosmo's highly proficient listening skills. These dogs are listening attentively.

Having witnessed canine skills, we humans are invited to 'become' dogs temporarily. I volunteer to leave the room whilst the other human researchers agree a task for me to complete. When I return, I listen out for, and respond to, the clicker – using its presence and its absence to guide my footsteps and actions, aiming to succeed in completing the unknown task.[7]

A tentative step
Click
Another tentative step
Noclick
Change direction and step tentatively
Noclick
Change direction and step tentatively
Click
A tentative step forward in the same direction
Click
Keep going in that direction
Click
And another step forward
Click
(If I had a tail, it might well be wagging now.)

The click becomes my focus. My movement, my embodiment, is an attending one that is aware of its attending, my ear seeming to cock and stretch into the space. As I work towards achieving what is required of me, albeit not as proficiently or rapidly as Winnie or Cosmo, I *feel* myself listening.

Click

My attention to the embodied act of attending as a practice of stretching invokes philosopher's Jean-Luc Nancy's thoughts on listening: 'to listen is *tender l'oreille* – literally, to stretch the ear – an expression that evokes a singular mobility, among the sensory apparatuses, of the pinna of the ear – it is an intensification and a concern, a curiosity or an anxiety' (2007, p. 5). I am particularly taken, in the context of co-production and participatory methods, with Nancy's multiple signification of listening as *concern*, *curiosity* and *anxiety* (2007, p. 5). In the current climate, I propose that we need more concern, curiosity and anxiety; that is, more of a certain kind of listening.

Click

A scene of contrast

We humans are sitting in a circle, on tree logs, in the middle of a small area of the Forest of Dean, surrounding a fire lit in the middle. Surrounding us, in turn, are Douglas Firs. It is the second day of the third workshop of *In conversation with. . . .* This one, facilitated by Richard Coles, Professor of Urban Landscape and Environmental Interaction at Birmingham City University, aims to explore whether and how arts-based methods, as forms of affective engagement, offer an alternative to the anthropocentric notions of conversation which prioritise vocal exchange.[8] We have been joined not only by trees for this workshop, but by two members of the Wye Valley Area of Outstanding Natural Beauty, Sarah Sawyer, Community Links Officer and Nikki Moore, Information Officer. Moore is also the Director of InsideOUT, a community organisation aiming to widen access to the Wye Valley.

The previous day, instructed by craftsman Dave Jackson of Wildwood Coppice Crafts, we each attempted to carve a wooden spoon out of a block of cherry tree wood. Sitting in a circle then too, on our tree logs, we were, for the most part, intently (and silently) focused on the task in hand. Later, facilitated by Sawyer and Moore, we walked through the forest decorating balls of clay with materials gathered to create 'bio-diversity balls', improvised scripts to imagine different trees' personalities and built miniature shelters on the forest floor.[9] At one point during the workshop, Sawyer invited each of us to go off and find a space in the forest and to sit there quietly on our own for five minutes. We vocally requested longer and, in the event, stretched it to more than 15.

Lying on the forest floor, I felt myself stretch again into an attentive listening:

breath
leaves rustling
branches creaking, squeaking
jet engine
footsteps
patter of rain on leaves
pattern of rain on leaves
rhythms
timbre, pitch
chainsaw
I am lying still
but my ears are stretched

Click

Now, back in our circle on the tree logs, Coles is asking us to reflect on our experiences of conversing with trees. In this moment of feedback we generate something that resembles audio-feedback – a loop between an input and output resulting in a deafening amplification. People are talked over, interrupted, dismissed, ignored, denied, marginalised and silenced. I feel dismayed and perplexed by our demonstrably poor communication. More specifically, I wonder how we can hope to engage more-than-human others as co-participants when we are incapable even – still – of listening to each human other? Is the shift towards exploring the more-than-human symptomatic of the failure of, or exhaustion with, the challenge of human-to-human relationships? I try to express the anxiety I have been feeling during the workshop: were we compelling things to talk to us, rather than *changing ourselves* so we could listen differently and more attentively? Philosopher and political theorist Nikolas Kompridis's question reveals the challenge: 'Should democratic politics require work on ourselves?' (2011, p. 255).

Sitting in our clearing in the forest, we seem a long way from the attentive listening of the dogs and our attentive listening as dogs, as well as our more recent attentive listening in the forest. We have unwittingly returned to familiar form.

Setting up conversations

Our research project, *In conversation with. . .: co-designing with more than human communities*, placed its emphasis on the co-designing aspect of participatory research. However, as the title suggests, there was an implicit presumption that conversation was a key part of the process, with the terms and practices attached to 'conversation' addressed through each workshop's design. How might we converse with dogs, trees, water and bees? What forms might conversation take and what might a conversation with these co-participants 'sound' like? What would approaching conversation as an expanded field of practice permit? Is conversation even the right mode or word?

Our conversational attempts with more-than-human communities are nested within a fairly well-established, but differentiated, set of critical discourses and approaches prompted by environmental crisis. Philosopher of science, Bruno Latour, for example, working at the level of epistemology, brings more-than-humans (in)to speech through a deconstructive move on what he proposes as the dichotomous realms of politics and nature – or the House of Humans/People and the House of Nature/Things. In the first house (categorised as epistemological) is 'the totality of speaking humans' who agree 'by convention to create fictions devoid of any external reality', whilst in the second house (categorised as ontological) are 'real objects' which 'have the property of defining what exists but lack the gift of speech' (2004, p. 14). The 'chattering of fictions' is thus in opposition to 'the silence of reality' (2004, p. 14).

Latour's critical turn rests in dissolving this distinction between humans who supposedly speak and nonhumans who are supposedly silent, through insisting on a shared 'speech impedimenta' and thus a shared need for speech prosthesis: 'I do not claim that things speak "on their own," since no beings, not even humans, speak on their own, but always *through something or someone else*' (2004, p. 68). Given the impossibility of facts speaking for themselves, Latour draws our attention to the presence of intermediaries or spokespersons and in highlighting these figures and their functions is able to pronounce that 'speech is no longer a specifically human property, or at least humans are no longer its sole masters' (2004, p. 65). In Latour's analytical frame, oriented towards and by Western science/political theory, it is through the speech prosthesis invented by 'lab coats' that nonhumans are enabled to '*participate in the discussions of humans, when humans become perplexed about the participation of new entities in political life*' (2004, p. 67, emphasis in original). The common world, for Latour, is an assemblage of humans and nonhumans, where 'matters of fact' (2004, p. 52), give way to 'matters of concern' (2004, p. 22), and 'disputed states of affair' (2004, p. 25). Introduced as matters of concern rather than matters of fact, nonhumans 'provoke perplexity and thus speech in those who gather around them, discuss them, and argue over them' (2004, p. 66). Different entities of the collective are thus stirred together 'in order to make them articulable and to *make them speak*' (2004, p. 89, emphasis in original).

Latour's approach diminishes human exceptionalism by insisting on the speech prosthesis common to all acts of speaking. Environmental feminist philosopher Val Plumwood also seeks to target ideas of human exceptionalism and the hubris generated. However, she does so by proposing that other species be recognised as communicative beings, rather than as 'simpler and lesser' subjects (2002, p. 189). What is required, she proposes, are dialogical methods of scientific practice 'which treat nature as active in the production of knowledge', the engagement founded on respect (2002, pp. 55–56). As Plumwood notes, most humans *have* had communicative experiences with animals. The task, then, is to communicate with other species including but beyond animals, and in such a way that such cross-species communication does not insist that other species learn a particular human communication method. For Plumwood, there is an ethics at stake in communicating with 'other species on their terms' (2002, pp. 55–56):

> The first thing we are likely to need in our philosophical toolkit is a communicative ethic, for we will need to ask permission to enter the interspecies community, and once inside we will need such an ethic to pursue negotiation and participation in a dialogical relationship with the other inhabitants.
>
> (2002, p. 188)

Important to my reflection on *In conversation with. . .*, is Plumwood's use of the phrase 'dialogical relationship', the foundation to which is an openness to the agency and communicative potential of other beings. Plumwood notes specifically 'sensitive listening', 'attentive observation' and 'an open stance that has not already closed itself off by stereotyping the other that is studied [. . .] as mindless and voiceless' (2002, p. 56). She offers a list of 'communicative virtues' intended to militate against hegemonic, human-centred stances of 'reduction, superiority and scepticism' (2002, p. 194). These virtues include recognising continuity-with rather than dualism-between, acknowledging humans' animality at the same time as species' differences, being open to the nonhuman other as 'an intentional and communicative being', being attentive, inviting communication, redistributing participation, resisting ranking, 'studying up', negotiating and admitting the other's complexity – the other's 'outside' of our knowledge (2002, p. 194).

In considering conversations with more-than-humans, other critics to reference briefly are political scientist Jane Bennett and feminist philosopher Donna Haraway. Bennett's recent work on the vitalism of materiality admits all 'things' as agential and interacting subjects capable of forming situations of concern – a participation in the world independent of human language which in turn challenges the prioritisation of such language as the gateway to participation. Agential things – forces and energies as much as matter – resonate beyond a semiotic or indexical reading. As Bennett writes: 'A vital materiality theory of democracy seeks to transform the divide between speaking subjects and mute objects into a set of differential tendencies and variable capacities' (2010, p. 108). In addressing distributed agencies, a 'confederation of human and non-human elements' (2010, p. x), where human and nonhuman are enmeshed, Bennett hopes to promote 'more attentive encounters between people-materialities and thing-materialities', 'giving a voice to a thing-power' (2010, p. 2). For Bennett, the ethical task is to become perceptually open to nonhuman vitality (2010, p. 14). Key words for me here are 'attentive', 'cultivation' and 'perceptually open'.[10]

The knots of Bennett's confederations bear a relation to Haraway's knotted encounters between companion species, though the knots of the latter are bound by relationships of regard, response and respect. In her introduction to *When Species Meet*, presenting the careful and committed work of scientist and bio-anthropologist Barbara Smuts, Haraway discusses nonlinguistic embodied communication, presented as 'more like a dance than a word', a 'co-constitutive naturalcultural dancing', 'the dance of "becoming with"' (2008, pp. 26–27). A core virtue expressed by Haraway – not least in her critique of Jacques Derrida's relationship with his cat – is curiosity. Lack of curiosity proposes a closed world (view), one that forecloses other worlds – a point related to Richard Sennett's sentiments that 'curiosity can "hearten" us to look beyond ourselves' (2012, p. 278). Haraway writes that whilst 'Derrida is relentlessly attentive to and humble before what he does not know' in his incuriosity towards his cat 'he missed a possible invitation, a possible introduction to other-worlding' (2008, p. 20).

Click

Listening again to Plumwood, Bennett and Haraway, a guide to participatory research might include the following checklist which would serve to remind human researchers of an ethical communicative stance, one appropriate to both humans and nonhumans: attentive, open, invitational, redistributive, informed, responsive, respectful, curious and humble. *Clicklist* might be a useful alternative word: a list that recalls us to appropriate attention, to a listening praxis.

Turn about with

The word 'conversation,' appearing in the 14th century, comes from the Latin, '*conversationem*', meaning 'act of living with', emerging from '*conversari*' – 'to live with, keep company with' or, more literally, 'to turn about with' (from *com* and *vertare*) (Dictionary.com, n.d.). This etymology recalls Haraway's species' companionship and the dance of communication. Following the line offered by turning about with invokes shifting, fluid perspectives created by companioning. This energy of turning about with brings me in turn to Gilles Deleuze's reflection on conversation being an 'active and creative line of flight' (2007, pp. 2, 10), generating a proliferation of unintended shoots through the dynamism afforded by a combination of at least two. One is always in conversation 'with'.

The value that Deleuze affords to two or more is common-place in my field of expertise – devised and collaborative performance. In devised performance practice, the performance is generated through a process of collaboration. Most typically, no script or text exists in advance of the collaboration, or any sense of an intended or expected outcome (Heddon and Milling 2005). The method is entirely process. Developing work in this manner, without a pre-existing play-text, requires the cultivation of certain dispositions, including awareness, concentration and trust (Oddey 1994, p. 173).[11]

A cardinal rule of devising practice is to keep the creative space open by remaining open to offers, resisting the impulse to refuse, reject or say no – blocks which block ideas. A simple workshop exercise, 'Yes, and. . .', demonstrates an attempt to develop a practice of openness: In pairs, one person makes an offer, and their partner must always add to it, beginning their contribution each time with 'Yes, and'. The story is allowed to continuously build over time; the end result is a collaboration – an example of co-authorship where each participant is necessary to the outcome and neither can claim origin or genius. Each offer is changed by the one that follows it; each offer is dependent on the one that precedes it. The story's infinite possibilities are engendered by the simple act of taking turns and being receptive to what is offered.

A: It's raining
B: Yes, and I've got a large umbrella
A: Let's shelter under it
B: Yes, and the wind is blowing us into the air
A: We are flying over the sea
B: Yes, and we have landed on an island.

(Farmer, n.d.)

In this exercise, no-one is able to imagine or control where the story will go, or where it will eventually end up. In this sense, the story remains always open, its future undetermined by what is already known. This improvised conversation reminds me of Kompridis's stress on the necessity for, and politics of, undetermined futures (2011, pp. 256–257), an approach which puts hubris into abeyance. Such an approach discloses possibilities not yet imagined.

One of the best-known devising companies, Chicago-based Goat Island, collaboratively produced a 'Letter to a Young Practitioner' – another example of co-authorship – which offers nuanced and evocative insight into the process of collaborative participation.[12] I encourage the reader to visit this letter and read it in full, but in the context of this chapter, some lines reverberate profoundly:

> Use what is around you, approach it with fresh eyes and ears: use the other workshop participants, Goat Island, the room you're in, the building, the city – other bodies. [. . .] Give up what seems important to you; it's not yours. [. . .] Accumulat[e] material, finding unexpected connections. [. . .] Collaborat[e] through words, sounds, touch, texture, viewing, thinking. The material is there to be received, processed, transformed. [. . . F]ocus on others to get out of your self. [. . .] It is shallow to rely on your own energy. Ideas like to be cross fertilized. [. . .] Beware of Brilliance. [. . .] Look for conflict resolution skills, forgiveness, the ability to listen, the ability to place faith in other people's fragmented ideas. [. . .] Never take the same route, always vary your path. [. . .] See as a new eye, as a novice, as someone who isn't jaded by fixed notions. [. . .] In an exercise on departure during a Goat Island workshop last summer, I was given a white sheet of paper from a participant with a single word written on it. The word was openness. [. . .] Taking forward the information given. This idea of ownership becomes a wider participation, and one of interaction and creativity with others. [. . .] Be confident and allow the material to come to you, begin to see with different eyes and learn the value of listening, the silence of yourself and others. [. . .] Be open to new discoveries. [. . .] Never think yourself singular. [. . .] Don't labor under the burden of importance.
>
> (Goat Island 2002)

It seems clear in Goat Island's work that the collaborators are both human and more-than-human: the room, the building, the city, the other bodies and the interactions between. The words used by Goat Island carry some of the learning that collaborative practice has taught me over the years, including openness, listening, discovery, creativity, receptivity, multiplicity and letting go. Listening again to Plumwood, Bennett and Haraway, the intersections are striking.

Click

Listening again

> *I am sitting in a circle, on tree logs, in the middle of a small area of the Forest of Dean. My human companions sit in the circle too, surrounding a fire lit in the middle. Surrounding us, in turn, are Douglas Firs. It is the second day of the third workshop of* In conversation with. . .*. Coles is asking us to reflect on*

our experiences of conversing with trees. In this moment of feedback we generate something that resembles audio-feedback – a loop between an input and output that results in a deafening amplification. People are talked over, interrupted, dismissed, ignored, denied, marginalised, silenced.

Though there is no space here to offer detailed speech analysis of the human conversation, let me place myself in this scene actively, because I am not an audience member or even witness, but a participant. In fact, and uninvited, I am the first to speak during the reflective feedback session. Throughout the 'conversation' I interrupt others, finish people's sentences, break the conversational thread by taking the discussion in unrelated directions and raise my voice over other speakers.

Listening to the audio recording of the conversation, it is clear that the distribution of noise between humans is not undifferentiated. Some people – those outside of academia or still training to be academics – talk less, are talked over more than others and are heard less; some people – mostly the male academics in our group – talk more than others and more frequently use phrases that signal either their 'mastery' over, or the self-evidence of, a subject (which comes down to the same thing); some people – mostly the female academics – ask more questions, apologise for interrupting and extend specific invitations to others to contribute. The academic speakers all tend to presume or insist on a shared foundation of knowledge across the group's human membership, thus ignoring the diverse make-up of the group (including academic diversity). Phrases such as 'we all know', and reference to particular theories through the use of shorthand terms, are frequent.[13]

In our staging of this conversation in the forest, we fail to attend to its, and our, dynamics and the effects – not to mention affects – of our conversational modalities. We fail to attend to each other, to what is unsaid and to what we might not yet know, as much as to what is actually spoken aloud. Talking here is demonstrably a stratified and targeted activity. The affective experiences of the past two days, stimulated through different arts-based exercises and intended to attune us differently, seem to have gone up in smoke as the majority of participants revert back to what might be deemed habitual practices in certain – most? – academic contexts. Though sitting in a clearing, and with a diverse group of participants (human and nonhuman alike), there is little that resembles an opening. In fact, the clearing has become the stage for the performance of authoritative knowledge and, importantly, particular kinds of knowledge. The space is filled with statements of certain fact rather than subjunctive proposition (Latour 2004, Sennett 2012).

Whilst the clearing in the forest, and the arrangement of seats in a circle, suggest a space open to all equally, in practice, dominant models of deliberation persist. Certain forms of speech – demonstrating 'rationale' thinking – are privileged. These seats in a circle resemble another, this one laid out by feminist scholar Iris Marion Young:

> So we have a different circle: Where there are structural inequalities of wealth and power, formally democratic procedures are likely to reinforce

> them, because privileged people are able to marginalize the voices and issues of those less privileged.
>
> (2000, p. 34)

Though class-based stratification is not so evident in our forest, privilege and power certainly are, with authority in the context of this 'research' workshop enacted through the normative performance of academic knowledge, serving to reinforce hierarchies and exclusions – human over human as well as human over nature. Dwelling on – attending to – this dismal stratification of human speaking is important. As Plumwood reminds us, the stakes attached to the redistribution of communication are profound: 'the kind of society whose democratic forms open communication and spread decision-making processes as equally as possible should, other things being equal, offer the best chance of effective action on these [. . .] ecoharms' (2002, p. 91). Sitting around a clearing in a forest is not enough. I wonder how and where the conversation would have flowed if we had insisted on saying 'Yes and' after every contribution?

> A distant sound is heard. It seems to come from the sky and is the sound of a breaking string. It dies away sadly. Silence follows, broken only by the thud of an axe striking a tree far away in the orchard.
>
> (Chekhov 1904, Act Four)

Con-versing

Since we are in the clearing in the forest, a sound from Heidegger is perhaps inevitable:

> It [. . .] might well be helpful to us to rid ourselves of only hearing what we already understand.
>
> (1971 cited Fiumara 1990, p. 38)

Hearing only what we already know might well be an example of the 'brilliance' Goat Island warns against. Though we aimed, in this project, to converse with more-than-human others, what we think we already know proved to be a hindrance to conversation. As Fiumara (1990, p. 106) so astutely puts it, our explicatory theories seem to seize and silence all objects; the moment they appear to speak to us through our discursive frames – the moment they are spoken for – is the moment we lose interest in them. I would propose that this is the moment, too, that they are rendered object rather than subject. This silencing of subjects through our explicatory theories resembles Plumwood's recognition of (human) language's potential to kill the vitality of subjects, with devastating consequences:

> A time-tested strategy for projects of mastery is the normalisation and enforcement of impoverishing, passifying and deadening vocabularies for what is to be reduced and ruthlessly consumed.
>
> (2002, p. 56)

Feminist environmental philosopher Catriona Sandilands asks similarly that, instead of (re)pressing the more-than-human through the frame of the discursive, the 'strangeness' of 'the human linguistic unknowability of nature' be preserved and fostered (1999, p. 185). Such 'strangeness', she argues, is foundational to the dynamics of conversation. For Sandilands – as for Plumwood – not only does such unknowability prompt human humility, it also keeps the conversation open and ongoing. In Sandilands' memorable words, 'Human language about nonhuman nature can never be complete, only by acknowledging its limits is the space opened for otherworldly conversations' (1999, p. 185).

Sandilands's proposal overlaps with my concern of compelling things to speak, a concern borrowed from theatre scholar Una Chaudhuri who, in turn, paraphrases cultural theorist Jean Baudrillard. Chaudhuri writes that the apparent 'silence' of animals 'seems to be able to survive all of the many ways humanity has tried to render them discursive' (2012, p. 55). Reading through Baudrillard, Chaudhuri proposes this silence as 'their continuing gift to us' (2012, p. 55). For Baudrillard:

> In a world bent on doing nothing but making one speak, in a world assembled under the hegemony of signs and discourse, their silence weighs more and more heavily on our organization of meaning.
>
> (1994 cited in Chaudhuri 2012, p. 55)

The silence that Baudrillard invokes is arguably a silence which results from the refusal to speak in terms already known. Silence is the speech of unknowness. Fiumara proposes that what is missing in this privileging of logos as assertive statement – 'the assertion of something about something' – is listening (1990, p. 3). There can be no speaking, however it is characterised, without listening; and yet, in Fiumara's view, speaking without listening 'has multiplied and spread, to finally constitute itself as a generalized form of domination and control' (1990, p. 2). Fiumara's critique of 'thinking-speaking' (1990, p. 13), directed primarily towards dialectical methods which dismantle only in order 'to repropose what has been demolished' (1990, p. 85), offers an accurate depiction of the scene in the forest, genuine dialogue replaced with competing monologues. The consequences of these habits of speaking-not-listening are far-reaching, with environmental deterioration 'probably [. . .] connected both to our benumbment and to cultural saturation with competitive expressions' (1990, p. 114).

Political theorist Andrew Dobson (2014) addresses this lacuna of listening in discussions of democratic practices (where the focus has tended to be on bringing things to speech), offering apophatic listening as the ideal form. In contrast with 'interruptive' listening, which *anticipates* the speaker's content and looks forward to contributing a sympathetic tale, apophatic listening is open and seeks to disclose rather than discover (Wolvin and Oakley 1996 cited Dobson 2014, p. 52). Apophatic listening resists too the monological risks of compassionate and sympathetic listening – both immersive forms of listening resulting in the production of a single perspective which mirrors back to the speaker exactly what he or she said, thereby diminishing the prospect of dialogue and mutual transformation.

Finally, apophatic listening contrasts with cataphatic listening, where the listener organises what is said through her or his own categories, learning nothing new and remaining unchanged by what is heard (Dobson 2014, pp. 67–68).

Dobson's proposition is indebted to the work of educationalist Leonard Waks, who presents apophatic listening as disinterested, unmotivated by any end point because accepting 'being with the other' as its own end (Waks 2007, p. 154). Unexpectedly, listening to Waks, I find myself back at the roots of conversation, *com-vertare*: to live with, keep company with, to turn about with. Here, also re-turning, are Haraway's dance of communication, Deleuze's 'lines of flight' and Goat Island's 'never singular' – dances, flights and collaborations are dependent on at least two participants, and not necessarily only human ones. In the act and practice of listening-speaking, meaning is co-authored, unexpected, arising from a *curious and attentive encounter* – between one and an other – rather than a retreat to the familiar and known (Sennett 2012, p. 21). Bringing Plumwood, Haraway, Sandihal, Fiumara and others into our forest clearing, it becomes clear to me that we need to be much better at listening others – human and nonhuman – to different forms of speech, including the con-verse, a versing with. The con-verse might well be a form of poetry that depends on at least two and welcomes the improvisational dance of unsettledness, uncertainty, provisionality and creativity, arising from the taking turns and turning-with-turning-with. Yes, and . . .. Such conversing could perform a radical and deeply necessary role in reconceptualising speaking, allowing it to be rendered otherwise, otherworldly and otherworlding.

Click Click Click

> You kick my imagination into the air like a particle of dust and it floats. But it's airborne with your imagination. Eventually, the two settle together on the floor, indistinguishable.
>
> (Goat Island 2002)

Click

> Yes, and ...
>
> You, a particle of dust, kick my imagination into the air, where we float together, airborne. Eventually, the two settle together on the floor, indistinguishable, both changed.
>
> (Goat Island 2002)

Notes

1 PAR offers a particular form of participatory research, distinct from other methods (e.g. participatory design or participatory ethnography).

2 This essay complements another (Heddon 2016). 'The cultivation of an entangled listening' was also prompted by my experience as a participant in the *In conversation*

with. . . workshops. That essay offers a different approach to the problem of listening, setting Jean-Luc Nancy's evocation of 'listening' (2007), alongside artist Adrian Howells's intimate performance works, in order to propose a performance practice that uses listening as a way to attend to the more-than-human. Read together, these two essays offer another conversation.

3 See http://www.ahrc.ac.uk/research/fundedthemesandprogrammes/crosscouncilprogrammes/connectedcommunities/ [accessed 15 May 2016]

4 Academic co-researchers were Michelle Bastian, Phil Jones, Richard Coles, Owain Jones, Deirdre Heddon, Martin Phillips, Julian Brigstocke, Johan Siebers, Niamh Moore and Clara Mancini. Other human participants included Tim Collins and Reiko Goto, Antony Lyons, members of the Evesham Beekeepers Association and workers with the Wye Valley Area of Outstanding Natural Beauty.

5 Note that, as of 2015, Dogs for the Disabled have changed their name to Dogs for Good.

6 See http://crc.open.ac.uk/Themes/ACI [accessed 15 May 2016]

7 For a short video recording of the workshop, which shows my attempt at becoming a dog, see https://www.youtube.com/watch?v=-dp1AnFIYH4#t=80 [accessed 15 May 2016]

8 See http://www.morethanhumanresearch.com/conversations-with-plants.html [accessed 15 May 2016]

9 These activities are the same ones that Sawyer and Moore use with a range of community groups who visit and use the site.

10 I recognise that my ongoing identification of such 'keynotes' and resonances – noted in places as speaking to the priorities of this chapter – risks paradox. Am I hearing only that which I already know? Am I surprised where I end up?

11 Students of devised theatre are typically introduced to collaborative-making skills through workshop practice. A frequently cited practitioner is Augusto Boal. As part of his 'Theatre of the Oppressed' portfolio, Boal (1992) formulated many workshop games and exercises intended to develop self-awareness and self-reflexivity in relation to practices and habits of participation (in life as well as in theatre), and to cultivate new skills, for example, attentiveness, observation, trust and team work.

12 Goat Island was founded in 1987 and performed its last work in 2009. See http://www.goatislandperformance.org/home.htm [accessed 15 May 2016].

In this letter, company members cite in turn from Virginia Woolf's 'A Letter to a Young Poet' and Rainer Maria Rilke's 'Letters to a Young Poet'.

13 Though this description is based on listening to the audio recording of the discussion, it is nevertheless impressionistic. The conversation – like most – is best described as a flow, with some people speaking more than others at certain times. For example, in some moments, the women in the group are more vocal; in the second half of the conversation, one of the male academics barely contributes and is invited at one point by one of the female academics to give his thoughts. There were also more male academic participants than female ones.

References

Bastian, M., 2013. *About* [online]. Available from: http://www.morethanhumanresearch.com/about.html [Accessed 15 May 2016].

Bastian, M., 2013. *More-than-human participatory research* [online]. Available from: http://www.morethanhumanresearch.com/ [Accessed 6 January 2016.]

Bennett, J., 2010. *Vibrant matter: a political ecology of things*. Durham and London: Duke University Press.

Boal, A., 1992. *Games for actors and non-actors*. London: Routledge.

Chaudhuri, U., 2012. The silence of the polar bears: performing (climate) change in the theater of species. *In*: W. Arons and T.J. May, eds. *Readings in performance and ecology*. New York: Palgrave Macmillan, 45–57.

Chekhov, A., 1904 (1998). The cherry orchard. *In*: *Anton Chekhov: five plays*. Trans. R. Hingley. Oxford: Oxford University Press.

Deleuze, G., and Parnet, C., 2007. *Dialogues II*. New York: Columbia University Press.

Dictionary.com, n.d. *Conversation* [online]. Available from: http://dictionary.reference.com/browse/conversation [accessed 15 May 2016].

Dobson, A., 2014. *Listening for democracy: recognition, representation, reconciliation*. Oxford: Oxford University Press.

Farmer, D., n.d. *Devising theatre* [online]. Available from: http://dramaresource.com/devising-theatre/ [accessed 3 May 2016].

Fiumara, G. C., 1990. *The other side of language: a philosophy of listening*. Trans. C. Lambert. London, New York: Routledge.

Goat Island, 2002. *Letter to a young poet* [online]. Available from: http://www.goatislandperformance.org/writing_L2YP.htm [accessed 6 January 2015].

Haraway, D., 2008. *When species meet*. Minneapolis, MN, London: University of Minnesota Press.

Heddon, D., 2013. *Co-design with dogs: meaningful participation and the problem with buttons* [online]. Available from: http://www.morethanhumanresearch.com/home/co-design-with-dogs-heddon [accessed 15 May 2016].

Heddon, D., forthcoming. The cultivation of entangled listening: an ensemble of more-than-human participants. *In*: A. Harpin and H. Nicholson, eds. *Performance and participation: practices, audiences, politics*. London: Palgrave Macmillan.

Heddon, D., and Milling, J., 2005. *Devising performance: a contemporary history*. London: Palgrave Macmillan.

Kompridis, N., 2011. Receptivity, politics, and democratic politics. *Ethics & Global Politics*, 4 (4), 255–272.

Latour, B., 2004. *Politics of nature: how to bring the sciences into democracy*. Trans. C. Porter. Cambridge, MA: Harvard University Press.

Nancy, J. L., 2007. *Listening*. Trans. C. Mandell. New York: Fordham University Press.

Oddey, A., 1994. *Devising theatre: a practical and theoretical handbook*. London: Routledge.

Pain, R., Whitman, G., and Milledge, D., 2010–2011. *Participatory action research toolkit: an introduction to using PAR as an approach to learning, research and action* [online] https://www.dur.ac.uk/resources/beacon/PARtoolkit.pdf [accessed 3 May 2016].

Plumwood, V., 2002. *Environmental culture: the ecological crisis of reason*. London: Routledge.

Sandilands, C., 1999. *The good-natured feminist: ecofeminism and the quest for democracy*. Minneapolis, MN: University of Minnesota Press.

Sennett, R., 2012. *The rituals, pleasures and politics of cooperation*. London: Allen Lane.

Waks, L., 2007. Listening and questioning: the apophatic/cataphatic distinction revisited. *Learning Inquiry*, 1 (2), 153–161.

Young, I. M., 2000. *Inclusion and democracy*. Oxford: Oxford University Press.

Index

Printed in the United States
by Baker & Taylor Publisher Services